HEATH
ALGEBRA 2
AN INTEGRATED APPROACH
LARSON, KANOLD, STIFF

RETEACHING
COPYMASTERS

Ann R. Kraus

McDougal Littell

Evanston, Illinois • Boston • Dallas

International Standard Book Number: 0-395-87201-4
1 2 3 4 5 6 7 8 9 10 HWI 01 00 99 98 97

What you should learn :

1.1	How to use the real number line to graph and order real numbers.

Correlation to Pupil's Textbook:

Mid-Chapter Test (p. 30)   **Chapter Test (p. 61)**
Exercises 1–4   Exercises 1, 2

Examples   *Graphing and Ordering Real Numbers*

a. Sketch the graph of the real numbers $-\frac{2}{7}$, $-\frac{4}{3}$, $\frac{12}{5}$, $\sqrt{3}$.

$-\frac{2}{7} \approx -0.29$, $-\frac{4}{3} \approx -1.33$, $\frac{12}{5} = 2.4$, $\sqrt{3} \approx 1.7$ *Write each number in decimal form.*

Draw a real number line. Plot the points.

b. Write the numbers in decreasing order: $-\sqrt{7}$, 0, $\frac{3}{2}$, -2, $\frac{3}{8}$.

$-\sqrt{7} \approx -2.6$, $\frac{3}{2} = 1.5$, $\frac{3}{8} \approx 0.38$ *Write each number in decimal form.*

Plot the points on a real number line.

$\frac{3}{2}$, $\frac{3}{8}$, 0, -2, $-\sqrt{7}$ *Write the numbers from largest to smallest.*

Guidelines: To order real numbers:
- Write each number in decimal form.
- Plot the points on a real number line.
- Observe the order of the numbers from your graph.

EXERCISES

In 1–12, write the numbers in increasing order.

1. 1, $\frac{1}{3}$, $\sqrt{2}$

2. $\frac{3}{5}$, -1, 1

3. $\sqrt{5}$, $\frac{2}{3}$, 3.25

4. -4, 1, $-\frac{5}{2}$

5. $\frac{3}{4}$, 2, $\frac{4}{3}$

6. 0, $-\sqrt{3}$, $-\frac{1}{2}$

7. 0, -2, $\frac{1}{3}$, $-\frac{2}{5}$, 1

8. $-\sqrt{2}$, -15, $\frac{1}{2}$, 5.7, 3

9. -1, $\sqrt{7}$, 2, $-\frac{1}{2}$, 0

10. $\frac{10}{3}$, -5, $\frac{5}{2}$, 1, $-\frac{5}{4}$

11. -0.3, $-\sqrt{3}$, $\sqrt{2}$, $\frac{1}{2}$, 4

12. 3, -5, 0, 4, $-\sqrt{5}$

Name _____

What you should learn :

Correlation to Pupil's Textbook:

Mid-Chapter Test (p. 30)   **Chapter Test (p. 61)**

Exercises 5–12, 17–19   Exercises 3–6, 17, 18

1.2	How to evaluate an algebraic expression.

Examples   *Evaluating Algebraic Expressions*

a. $2(3 + 18 \div 9 - 7) = 2(3 + 2 - 7)$ *Divide.*

$\qquad\qquad\qquad\qquad = 2(-2)$ *Add within parentheses.*

$\qquad\qquad\qquad\qquad = -4$ *Multiply.*

b. $2(-1 + 3) - 4^2 \div 8 = 2(2) - 4^2 \div 8$ *Add within parentheses.*

$\qquad\qquad\qquad\qquad = 2(2) - 16 \div 8$ *Evaluate the power.*

$\qquad\qquad\qquad\qquad = 4 - 2$ *Multiply and divide.*

$\qquad\qquad\qquad\qquad = 2$ *Subtract.*

c. Evaluate $2t^2 - 3$ when $t = 4$.

$2(t)^2 - 3 = 2(4)^2 - 3$ *Substitute 4 for t.*

$\qquad\quad = 2(16) - 3$ *Evaluate the power.*

$\qquad\quad = 32 - 3$ *Multiply.*

$\qquad\quad = 29$ *Subtract.*

Guidelines: To evaluate algebraic expressions:
- Substitute the given values into the algebraic expression.
- Follow the order of operations table on page 10 of the textbook.

EXERCISES

In 1–10, evaluate the expression.

1. $2a + 3$, when $a = 7$

2. $4x - 3y + 2$, when $x = 3$ and $y = 4$

3. $4x - 3y + 2$, when $x = 4$ and $y = 3$

4. $15p(4 - p)$, when $p = 2$

5. $3m - (m + 5)$, when $m = 20$

6. $(3x + 1) \div y$, when $x = 7$ and $y = 11$

7. $(2x)^2 - 3$, when $x = 4$

8. $3b^3 + 4$, when $b = 2$

9. $x^2 + y^2$, when $x = \frac{1}{2}$ and $y = 2$

10. $9(m - n)^3 + 5$, when $m = 4$ and $n = 1$

Scrambled answers for first column of exercises: 9, 35, 17, $\frac{17}{4}$, 61

Name _____

What you should learn :

| **1.3** | How to solve a linear equation. |

Correlation to Pupil's Textbook:

Mid-Chapter Test (p. 30)   **Chapter Test (p. 61)**

Exercises 13–16   Exercises 7, 8

| **Examples** | *Solving a Linear Equation* |

a. $4s - 6 = 7s + 3$ *Original equation*

$\qquad -6 = 3s + 3$ *To collect the variable terms, subtract $4s$ from both sides.*

$\qquad -9 = 3s$ *To isolate the variable term, subtract 3 from both sides.*

$\qquad -3 = s$ *Divide both sides by 3.*

b. $5(x - 3) + 9 = -2(x - 2)$ *Original equation*

$\quad 5x - 15 + 9 = -2x + 4$ *Distributive property*

$\qquad\quad 5x - 6 = -2x + 4$ *Simplify.*

$\qquad\qquad 7x - 6 = 4$ *Add $2x$ to both sides.*

$\qquad\qquad\quad 7x = 10$ *Add 6 to both sides.*

$\qquad\qquad\quad\ x = \frac{10}{7}$ *Divide both sides by 7.*

Guidelines: To solve a linear equation:
- Collect the variable terms on one side of the equation.
- Isolate the variable term.
- Divide both sides of the equation by the variable coefficient.
- Check the answer in the original equation.

EXERCISES

In 1–15, solve the equation.

1. $-\dfrac{x}{3} = 2$

2. $-19 = y + 5$

3. $9 - z = 5$

4. $4 + 6x = 12$

5. $\frac{1}{2}x - 5 = 1$

6. $3(5 - a) = -4(a - 4)$

7. $4x - 2x = 15 - 3x$

8. $3x + 9 = 2(x - 5)$

9. $-4(k - 2) + 3(k + 1) = 7$

10. $15(4 - y) = 5(10 + 2y)$

11. $-3(m - 2) = 6(4m + 1)$

12. $-2x = 2(x + 1) + 3$

13. $6n = \frac{2}{3}(5n - 2)$

14. $2(4t) - (t - 1) = 2(1 - t)$

15. $15(x - 2) = -13(1 + x) + 11$

Scrambled answers for first column of exercises: 12, 4, 1, -6, 4, $-\frac{1}{2}$, 0, 3

Reteach
Chapter 1

What you should learn :

1.4	How to use problem-solving strategies to solve problems.

Correlation to Pupil's Textbook:

Mid-Chapter Test (p. 30) **Chapter Test (p. 61)**
Exercise 20   Exercise 19

Example *Creating a Linear Model to Solve a Real-Life Problem*

Pete completes a 17.4 mile marathon in 3 hours by running some of the distance and walking the rest of the way. If he runs at the rate of 7 miles per hour and walks at the rate of 4 mile per hour, how long did Pete run?

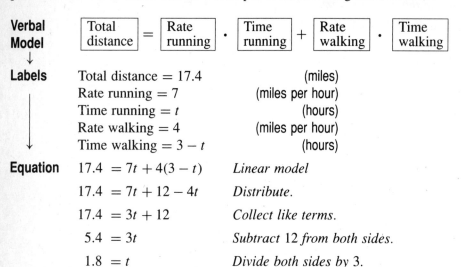

Verbal Model →

$$\boxed{\text{Total distance}} = \boxed{\text{Rate running}} \cdot \boxed{\text{Time running}} + \boxed{\text{Rate walking}} \cdot \boxed{\text{Time walking}}$$

Labels

Total distance = 17.4   (miles)
Rate running = 7   (miles per hour)
Time running = t (hours)
Rate walking = 4   (miles per hour)
Time walking = $3 - t$ (hours)

Equation

$17.4 = 7t + 4(3 - t)$ *Linear model*

$17.4 = 7t + 12 - 4t$ *Distribute.*

$17.4 = 3t + 12$ *Collect like terms.*

$5.4 = 3t$ *Subtract 12 from both sides.*

$1.8 = t$ *Divide both sides by 3.*

Pete ran for 1.8 hours.

Guidelines: To solve a word problem, use the following:

- Write a verbal model.
- Assign labels.
- Write an algebraic model.
- Solve the algebraic model.
- Answer the question.

EXERCISES

1. Jason must sell $100 worth of peat moss in order to go on a camping trip with his scout troop. The peat moss costs $2.50 per bag and he has already sold 12 bags. How many more bags of peat moss must Jason sell?

2. The cost for a long distance telephone call is $0.31 for the first minute and $0.24 for each additional minute (or portion thereof). If the total cost for the call is $4.87, how long did the call last?

Reteach

Chapter 1

Name _____

What you should learn :

1.5	How to solve a literal equation for a specific variable and how to evaluate it.

Correlation to Pupil's Textbook:

Chapter Test (p. 61)
Exercises 11, 12

Examples | *Solving a Literal Equation*

a. Solve for C.

$$F = \tfrac{9}{5}C + 32$$ *Formula for conversion from Celsius to Fahrenheit.*

$$F - 32 = \tfrac{9}{5}C$$ *To isolate $\tfrac{9}{5}C$, subtract 32 from both sides.*

$$\tfrac{5}{9}(F - 32) = C$$ *Multiply both sides by $\tfrac{5}{9}$.*

b. The formula for calculating the selling price is $S = L - rL$, where L is the list price and r is the rate of discount. An automatic coffee maker is advertised to sell for \$53.55 which is a 15% rate of discount. Calculate the list price.

$$S = L - rL$$ *Formula for selling price*

$$S = L(1 - r)$$ *Factor L out of the terms on the right side.*

$$\frac{S}{1 - r} = L$$ *Divide both sides by $1 - r$.*

$$\frac{53.55}{1 - 0.15} = L$$ *Substitute 53.55 for S and 0.15 for r.*

$$63 = L$$ *Simplify.*

The list price is \$63.00.

Guidelines: To solve a literal equation for a specific variable:
- Collect all terms with a specified variable on one side of the equation.
- Isolate the variable terms.
- If more than one term, factor out the specified variable.
- Divide both sides by the coefficient of the specified variable.

EXERCISES

In 1–8, solve for the indicated variable.

1. Solve for $r : A = 2\pi rh$

2. Solve for $w : V = lwh$

3. Solve for $b : \dfrac{b}{h} = \dfrac{3}{4}$

4. Solve for $h : V = \dfrac{\pi}{3}r^2 h$

5. Solve for $r : S = L - rL$

6. Solve for $t : A = P + Prt$

7. Solve for $x : p = 12 - \dfrac{x}{1000}$

8. Solve for $h : V = \tfrac{1}{3}b^2 h$

9. The cost is $C = 5000 + 0.56x$ where x is the number of items produced. Solve the equation for x. Then evaluate x when $C = 8360$.

(c) D.C. Heath and Company

Name _____

What you should learn :

| 1.6 | How to solve linear inequalities. |

Correlation to Pupil's Textbook:

Chapter Test (p. 61)
Exercises 13, 14, 21

| **Examples** | *Solving Linear Inequalities* |

a.

$3 - 2x \geq 5$	*Original inequality*
$-2x \geq 2$	*To isolate $-2x$, subtract 3 from both sides.*
$x \leq -1$	*Divide both sides by -2. (Multiplying or dividing by a negative number reverses the inequality sign.)*

 *Graph of the inequality*

b.

$-2 < 1 - 3x < 10$	*Original compound inequality*
$-3 < -3x < 9$	*To isolate $3x$, subtract 1 from each expression.*
$1 > x > -3$	*To solve for x, divide each expression by -3.*
$-3 < x < 1$	*Rewrite the inequality.*

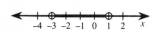 *Graph of the original inequality*

c. $2x - 5 \geq 1$ or $2x - 5 \leq -1$ Original compound inequality

Solution of first inequality: Solution of second inequality:

$2x - 5 \geq 1$	*First inequality.*	$2x - 5 \leq -1$	*Second inequality.*
$2x \geq 6$	*Add 5 to both sides.*	$2x \leq 4$	*Add 5 to both sides.*
$x \geq 3$	*Divide both sides by 2.*	$x \leq 2$	*Divide both sides by 2.*

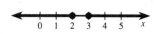 *Graph of the compound inequality*

Guidelines: To solve a linear inequality:
- Isolate the variable on one side of the inequality using transformations that produce an equivalent inequality as listed on page 37 of the textbook.

EXERCISES

In 1–9, solve the inequality. Then sketch its graph.

1. $5x - 2 < \frac{1}{2}$ **2.** $3x + 4 \geq -2$ **3.** $-2x - 1 > 5$

4. $3 - 4x \leq 1$ **5.** $-x - 1 < -1$ **6.** $3 < 2x - 1 < 7$

7. $2 \leq 3 - x \leq 8$ **8.** $3x - 2 \leq -5$ or $x - 1 \geq 0$ **9.** $-2 < \frac{1}{2}x + 1 < 3$

Scrambled answers for the first column of exercises: $-5 \leq x \leq 1$, $x < \frac{1}{2}$, $x \geq \frac{1}{2}$

Reteach
Chapter 1

What you should learn :

1.7	How to solve absolute value equations and inequalities.

Correlation to Pupil's Textbook:

Chapter Test (p. 61)
Exercises 9, 10, 15, 16

Examples   *Solving Absolute Value Equations and Inequalities*

a. $|3x + 1| = 5$ *Original equation*

     $3x + 1 = 5$ or $3x + 1 = -5$ *Transform original equation into two linear equations.*

       $3x = 4$ or $3x = -6$ *Subtract 1 from both sides.*

         $x = \frac{4}{3}$ or $x = -2$ *Divide both sides by 3.*

b. $|3 - 2x| \leq 7$ *Original inequality*

    $-7 < 3 - 2x \leq 7$ *Write equivalent compound inequality.*

  $-10 < \quad -2x \quad \leq 4$ *Subtract 3 from each expression.*

     $5 \geq \quad x \quad \geq -2$ *Divide each expression by -2.*

    $-2 \leq \quad x \quad \leq 5$ *Rewrite the inequality.*

c. $|x - 6| > 4$ *Original inequality*

   $x - 6 < -4$ or $x - 6 > 4$ *Write equivalent compound inequality.*

     $x < 2$ or $x > 10$ *Add 6 to both sides of each inequality.*

Guidelines: To solve absolute value equations or inequalities:
- Transform an absolute value equation into two linear equations.
- Transform an absolute value inequality into the equivalent compound inequality using the rules on page 44 of the textbook.
- Solve for the variable.

EXERCISES

In 1 and 2, solve the equation.

   1. $|5r - 8| = 2$ **2.** $|17m + 13| = 4$

In 3–6, solve the inequality.

   3. $|4x - 5| < 5$ **4.** $|2x + 1| \leq 3$

   5. $|y + 3| > 5$ **6.** $|\frac{1}{2}x - 2| \geq 7$

Name _____

What you should learn :

1.8	How to use tables and graphs to organize data.

Correlation to Pupil's Textbook:

Chapter Test (p. 61)

Exercise 20

Example	*Organizing Data*

At a car dealership, the number of new cars sold in a week by each
salesperson was as follows: 5, 8, 2, 0, 2, 4, 7, 4, 1, 1, 2, 2, 0, 1, 2, 0, 1, 3, 3, 2.

a. Construct a frequency distribution for this data.

Number	Tally	Frequency
8	|	1
7	|	1
6		0
5	|	1
4	||	2
3	||	2
2	ⵑ |	6
1	||||	4
0	|||	3

b. Construct a line plot for this data.

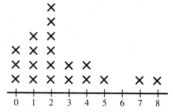

c. Construct a bar graph that shows the number of
salespeople who sold 0–8 cars.

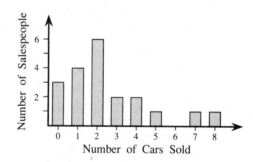

Guidelines: A collection of data can be organized by using a table, a frequency
distribution, a line plot, a bar graph, a circle graph, or a time-line graph.

EXERCISES

1. Twenty-eight students in a class were asked how many cars their family
owned. The results were as follows:

2, 2, 3, 2, 1, 2, 2, 4, 3, 2, 0, 1, 0, 1, 1, 2, 2, 3, 2, 3, 3, 5, 1, 1, 3, 0, 1, 2.

Constuct a frequency distribution and a line plot for this data.

2. Each of the members of a recent high school graduating
class was asked to name his/her favorite among these
subjects: English, foreign language, history, mathematics,
science. The results are shown in the table. Construct a
bar graph that shows these results.

English	62
Foreign language	40
History	40
Mathematics	18
Science	33

Name _____

What you should learn :

2.1	How to sketch a graph using a table of values or by identifying it as a horizontal or vertical line.

Correlation to Pupil's Textbook:

Mid-Chapter Test (p. 93)
Exercises 1, 2

Examples | *Sketching the Graph of an Equation*

a. Sketch the graph of $y = 2x - 2$.

First, construct a table of values.

x	-1	0	1	2	3
y	-4	-2	0	2	4

Then, plot the five points in the table. Finally, draw a line through the points.

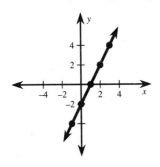

b. Sketch the graph of $x = 3$.

The graph is a vertical straight line. Every point on the line has an x-coordinate of 3. y can have any value.

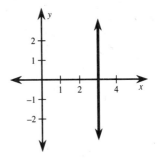

Guidelines: To sketch the graph of an equation:
- Decide whether the graph is a vertical line: $x = a$ or a horizontal line: $y = b$.
- Construct a table of values.
- Plot the points.
- Draw a line through the points.

EXERCISES

In 1–15, sketch the graph of the equation.

1. $y = 2x + 1$ **2.** $y = x - 5$ **3.** $y = 2$

4. $y = x + 1$ **5.** $y = 2x - 1$ **6.** $y = -x + 2$

7. $y = -x$ **8.** $x = 5$ **9.** $y = -1$

10. $y = x$ **11.** $y = -3x + 3$ **12.** $x = -2$

13. $y = 5x + 3$ **14.** $y = x + 6$ **15.** $y = -x - 1$

Name _____

What you should learn :

| 2.2 | How to find the slope of a line and identify parallel and perpendicular lines. |

Correlation to Pupil's Textbook:

| **Mid-Chapter Test (p. 93)** | **Chapter Test (p. 119)** |
| Exercises 3–8, 18, 20 | Exercises 1, 2, 7, 8, 20 |

| **Examples** | *Finding and Interpreting the Slope of a Line* |

a. Find the slope of the line containing $(2, -1)$ and $(4, -2)$.

Let $(x_1, y_1) = (2, -1)$ and let $(x_2, y_2) = (4, -2)$.

$$m = \frac{y_2 - y_1}{x_2 - x_1} \qquad m = \frac{Rise}{Run} = \frac{Difference\ in\ y\text{-}values}{Difference\ in\ x\text{-}values}$$

$$= \frac{-2 - (-1)}{4 - 2} \qquad \textit{Substitute values.}$$

$$= -\frac{1}{2} \qquad \textit{Simplify.}$$

b. Decide whether the two lines are parallel, perpendicular, or neither. Line 1 contains the points $(0, 0)$ and $(3, 2)$. Line 2 contains the points $(2, 0)$ and $(-1, -2)$.

$$m_1 = \frac{2 - 0}{3 - 0} = \frac{2}{3} \qquad \textit{Substitute the points for Line 1 into the slope formula.}$$

$$m_2 = \frac{-2 - 0}{-1 - 2} = \frac{2}{3} \qquad \textit{Substitute the points for Line 2 into the slope formula.}$$

Because the slopes are equal, Line 1 and Line 2 are parallel.

Guidelines:

To find the slope of a line:
- Represent the given points as (x_1, y_1) and (x_2, y_2).
- Calculate the rise, $y_2 - y_1$, and the run, $x_2 - x_1$.
- Divide the rise by the run.

To determine whether two different lines are parallel or perpendicular:
- Calculate the slope of each line.
- The lines are parallel if the slopes are equal.
- The lines are perpendicular if the slopes are negative reciprocals.

EXERCISES

In 1–8, find the slope of the line containing the two points.

1. $(1, 1), (3, 4)$ **2.** $(-1, 2), (2, 5)$ **3.** $(2, 4), (0, 5)$ **4.** $(3, -2), (3, 1)$

5. $(-3, -1), (-1, 1)$ **6.** $(2, 0), (0, 2)$ **7.** $(-4, -1), (1, -1)$ **8.** $(2, 1), (\frac{1}{2}, -2)$

In 9 and 10, decide whether the lines are parallel, perpendicular, or neither.

9. Line 1 contains $(-2, 6)$ and $(2, 7)$.
Line 2 contains $(4, -1)$ and $(5, -5)$.

10. Line 1 contains $(1, 2)$ and $(5, -1)$.
Line 2 contains $(-2, -2)$ and $(2, -1)$.

Scrambled answers for first row of Exercise 1–8: 1, undefined, $\frac{3}{2}$, $-\frac{1}{2}$

Reteach
Chapter 2

Name _____

What you should learn :

| 2.3 | How to sketch a quick graph of a line using intercepts or the slope-intercept form. |

Correlation to Pupil's Textbook:

Mid-Chapter Test (p. 93)   **Chapter Test (p. 119)**
Exercises 9–14   Exercises 3–6

Examples   *Sketching a Quick Graph of a Line*

a.

$3x - 4y = 12$	*Original equation*
$3x - 4(0) = 12$	*Let $y = 0$.*
$x = 4$	*The x-intercept is 4.*
$3(0) - 4y = 12$	*Let $x = 0$ in the original equation.*
$y = -3$	*The y-intercept is -3.*

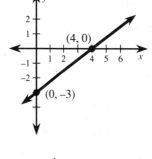

Plot the intercepts (4, 0) and (0 − 3). Then draw a line through the points to obtain the graph shown at the right.

b.

$2x + 3y = 3$	*Original equation*
$3y = -2x + 3$	*Subtract $2x$ from both sides.*
$y = \underbrace{-\frac{2}{3}}_{m} x + \underbrace{1}_{b}$	*Divide both sides by 3.*

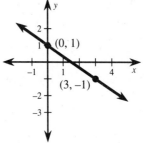

Plot the y-intercept (0, 1). To locate a second point, move 2 units down and 3 units to the right. Then draw a line through the points to obtain the graph shown at the right.

Guidelines: To sketch a quick graph of a line:
- Find the x-intercept and y-intercept.
- Plot the intercepts.
- Draw a line through the two points.

OR

- Write the equation in slope-intercept form.
- Plot the y-intercept.
- Use the slope to locate a second point.
- Draw a line through the two points.

EXERCISES

In 1–3, sketch the line. Label the coordinates of the x-intercept and y-intercept.

  1. $2x - y = 10$ **2.** $2x + 3y = 12$ **3.** $-3x + 4y = -2$

In 4–6, sketch the line using the slope and y-intercept.

  4. $3x + 4y = 4$ **5.** $3x - y = 2$ **6.** $-3x + 2y = -6$

Name _____

What you should learn :

| 2.4 | How to write equations of lines. |

Correlation to Pupil's Textbook:

Mid-Chapter Test (p. 93)   **Chapter Test (p. 119)**
Exercises 15–17, 19   Exercises 9, 10, 19

Examples | *Writing an Equation of a Line*

a. Write an equation of the line with y-intercept $(0, -2)$ and a slope of $\frac{1}{3}$.

$\quad y = mx + b$ *Slope-intercept form*

$\quad y = \frac{1}{3}x - 2$ *Substitute $\frac{1}{3}$ for m and -2 for b.*

b. Write an equation of the line containing the point $(-4, 1)$ with a slope of $-\frac{5}{2}$.

$\quad y - y_1 = m(x - x_1)$ *Point-slope form*

$\quad y - 1 = -\frac{5}{2}(x - (-4))$ *Substitute $-\frac{5}{2}$ for m, -4 for x_1, and 1 for y_1.*

$\quad y - 1 = -\frac{5}{2}(x + 4)$ *Simplify.*

$\quad y - 1 = -\frac{5}{2}x - 10$ *Distributive Property*

$\quad y = -\frac{5}{2}x - 9$ *Slope-intercept form*

c. Write an equation of the line containing the points $(-1, 1)$ and $(1, 5)$.

$\quad m = \dfrac{y_2 - y_1}{x_2 - x_1} = \dfrac{5 - 1}{1 - (-1)} = \dfrac{4}{2} = 2$ *Calculate the slope of the line.*

$\quad y - y_1 = m(x - x_1)$ *Point-slope form*

$\quad y - 1 = 2(x - (-1))$ *Substitute 2 for m, -1 for x_1, and 1 for y_1.*

$\quad y - 1 = 2(x + 1)$ *Simplify.*

$\quad y - 1 = 2x + 2$ *Distributive Property*

$\quad y = 2x + 3$ *Slope-intercept form*

Guidelines: To write an equation of a line:
- Calculate the slope (if not given).
- If the y-intercept is given, use slope-intercept form.
- If the y-intercept is not given, use point-slope form.

EXERCISES

In 1–6, write an equation of the line containing the given point with the given slope.

1. $(0, \frac{2}{3})$, $m = 4$ **2.** $(5, 3)$, $m = \frac{1}{2}$ **3.** $(1, 0)$, $m = -1$

4. $(2, -1)$, $m = \frac{2}{3}$ **5.** $(-2, 0)$, $m = -\frac{1}{3}$ **6.** $(0, -2)$, $m = -\frac{1}{4}$

In 7–9, write an equation of the line containing the two points.

7. $(-6, -1)$, $(3, 2)$ **8.** $(-4, 3)$, $(0, -5)$ **9.** $(-3, 3)$, $(3, -6)$

Scrambled answers for first column of exercises: $y = \frac{1}{3}x + 1$, $y = 4x + \frac{2}{3}$, $y = \frac{2}{3}x - \frac{7}{3}$

Reteach
Chapter 2

What you should learn :

2.5	How to graph a linear inequality in two variables.

Correlation to Pupil's Textbook:

Chapter Test (p. 119)
Exercises 11, 12, 15–18

Examples | *Sketching the Graph of a Linear Inequality*

a. Sketch the graph of the linear inequality $x < -2$.

Begin by sketching the graph of the vertical line $x = -2$. (Use a dashed line because the inequality is "$<$.")

Test $(0, \ 0)$ in the inequality. This point lies to the right of the vertical dashed line and it is *not* a solution.

The graph consists of *all* points to the left of the vertical dashed line, as shown at the right.

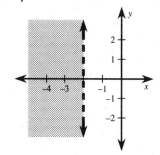

b. Sketch the graph of the linear inequality $3x - 5y \geq -15$.

Begin by sketching the graph of the line $3x - 5y = -15$. (Use a solid line because the inequality is "$\geq$.")

Test $(0, \ 0)$ in the inequality. This point lies below the line and it *is* a solution.

The graph consists of *all* points on or below the line, as shown at the right.

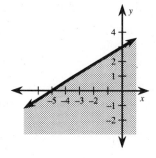

Guidelines: To sketch the graph of a linear inequality:
- Write the corresponding equation.
- Sketch the line given by the corresponding equation. (Use a dashed line for inequalities with $<$ or $>$ and a solid line for inequalities with $\leq$ or $\geq$.)
- Test one point (that is not on the line) in the inequality.
- If this point satisfies the inequality, all points in the half-plane containing the point are solutions of the inequality. Otherwise, the solution is the other half-plane.

EXERCISES

In 1–12, sketch the graph of the inequality.

1. $x > 3$

2. $-y > \frac{5}{2}$

3. $15x > -10$

4. $4x \leq \frac{4}{3}$

5. $x + y \leq 2$

6. $x - y > 1$

7. $x - 2y \geq 4$

8. $2x + 3y \geq -12$

9. $y < 2x + 1$

10. $y > \frac{3}{2}x - 3$

11. $x \geq y$

12. $-2x + 5y \geq 10$

Reteach
Chapter 2

What you should learn :

| 2.6 | How to graph absolute value equations. |

Correlation to Pupil's Textbook:

Chapter Test (p. 119)
Exercises 13, 14

Examples | *Sketching the Graph of an Absolute Value Equation*

a. Sketch the graph of $y = |2x - 3| - 2$.

Begin by setting the quantity inside the absolute value signs equal to zero and solving for x: the solution is $x = \frac{3}{2}$. This is the x-coordinate of the vertex. Now, construct a table of values.

x	−1	0	1	$\frac{3}{2}$	2	3	4
y	3	1	−1	−2	−1	1	3

Plot the points in the table. The vertex is at $(\frac{3}{2}, -2)$. Draw a $\bigvee$-shaped graph through the points, as shown at the right.

b. Sketch the graph of $y = -|x + 1| + 1$.

Begin by setting the quantity inside the absolute value signs equal to zero and solving for x: the solution is $x = -1$. This is the x-coordinate of the vertex. Now, construct a table of values.

x	−4	−3	−2	−1	0	1	2
y	−2	−1	0	1	0	−1	−2

Plot the points in the table. The vertex is at $(-1, 1)$. Draw a $\bigvee$-shaped graph through the points, as shown at the right.

Guidelines: To sketch the graph of $y = a|bx + c| + d$:
- Solve $bx + c = 0$ to find the x-coordinate of the vertex.
- Construct a table of values using the x-coordinate of the vertex, some x-values to the right, and some x-values to the left.
- Plot the points given in the table.
- Draw a $\bigvee$-shaped graph through the points.

EXERCISES

In 1–12, sketch the graph of the equation.

1. $y = |x - 2|$ **2.** $y = |x + 3|$ **3.** $y = -|x| + 4$ **4.** $y = -|x + 2| - 2$

5. $y = |x - 1| + 2$ **6.** $y = |2x - 4| - 1$ **7.** $y = |3 - x| + 5$ **8.** $y = -|2 - x|$

9. $y = |2x + 1|$ **10.** $y = |3x| - 2$ **11.** $y = |\frac{1}{2}x - 3|$ **12.** $y = -2|1 + 2x| + 3$

Reteach

Chapter 2

Name _____

Correlation to Pupil's Textbook:

Chapter Test (p. 119)
Exercise 21

What you should learn :

2.7	How to write an equation of the line that best fits a collection of data.

Example	*Fitting a Line to Data*

Approximate the best-fitting line for the data in the table.

x	1	1.5	2	3	4	5	6	7	7.5	8
y	7.5	6	6	5	4.5	5	3	3.5	4	3.5

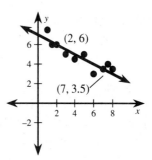

To begin, draw a scatter plot of the data. Then sketch the line that best fits the points, with about as many points above the line as below, as shown at the right.

Now, locate two points on the line and use these points to find an equation of the line.

$$m = \frac{3.5 - 6}{7 - 2} = \frac{-2.5}{5} = -\frac{1}{2}$$ *Calculate the slope of the line.*

$y - y_1 = m(x - x_1)$ *Point-slope form*

$y - 6 = -\frac{1}{2}(x - 2)$ *Substitute $(2, 6)$ for (x_1, y_1) and $-\frac{1}{2}$ for m.*

$y - 6 = -\frac{1}{2}x + 1$ *Distributive Property*

$y = -\frac{1}{2}x + 7$ *Solve for y.*

Guidelines: To fit a line to data:
- Follow the steps outlined on page 107 of the textbook.

EXERCISES

1. Approximate the best-fitting line for the data in the table.

x	−1	−0.5	0.5	1	1.5	2	3	3.5	4	4.2
y	8	8	7	5.5	10	3	3	0.5	0	−2

2. The data in the table shows the age, *t* (in years), and the corresponding height, *h* (in inches), for a young man from the age of 2 to the age of 19. Approximate the best-fitting line for this data.

Age (t)	2	3	6	8	10	12	14	15	17	18	19
Height (h)	28	33	40	46	52	55	61	64	70	72	72

Reteach
Chapter 3

Name _____

What you should learn :

| 3.1 | How to solve a system of linear equations in two variables by graphing. |

Correlation to Pupil's Textbook:

Mid-Chapter Test (p. 144)
Exercises 1–6, 13, 16

Examples | *Graphing and Solving a Linear System*

a. Solve the system. $\begin{cases} x - y = 1 & \text{Equation 1} \\ 2x + y = -4 & \text{Equation 2} \end{cases}$

Begin by writing each equation in slope-intercept form.

$$y = x - 1 \qquad \text{Equation 1}$$
$$y = -2x - 4 \qquad \text{Equation 2}$$

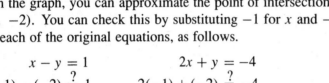

Then sketch a quick graph of each equation, as shown at the right. From the graph, you can approximate the point of intersection to be $(-1, -2)$. You can check this by substituting -1 for x and -2 for y into each of the original equations, as follows.

$$x - y = 1 \qquad\qquad 2x + y = -4$$
$$(-1) - (-2) \overset{?}{=} 1 \qquad 2(-1) + (-2) \overset{?}{=} -4$$
$$1 = 1 \qquad\qquad\qquad -4 = -4$$

b. Solve the system. $\begin{cases} x + 3y = 3 & \text{Equation 1} \\ x + 3y = -3 & \text{Equation 2} \end{cases}$

Begin by writing each equation in slope-intercept form.

$$y = -\tfrac{1}{3}x + 1 \qquad \text{Equation 1}$$
$$y = -\tfrac{1}{3}x - 1 \qquad \text{Equation 2}$$

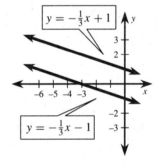

Then sketch a quick graph of each equation, as shown at the right. Note that the lines are parallel because they have the same slope. The lines have no point of intersection and the system has no solution.

Guidelines: To solve a system of linear equations by graphing:
- Write each equation in slope-intercept form.
- Sketch the graph of each line.
- From the graph, observe whether there is one solution, no solution, or infinitely many solutions.
- If there is a solution, check it algebraically.

EXERCISES

In 1–6, sketch the graph of the linear system. Then use the graph to estimate the solution. If there is exactly one solution, check it algebraically.

1. $\begin{cases} x - 2y = 4 \\ x + y = 4 \end{cases}$
2. $\begin{cases} 3x + y = 6 \\ 2x + 3y = -3 \end{cases}$
3. $\begin{cases} 2x - y = -4 \\ y = 3 \end{cases}$

4. $\begin{cases} 2x + 4y = 7 \\ 3x + 6y = 5 \end{cases}$
5. $\begin{cases} 6x - 8y = 2 \\ \tfrac{9}{2}x - 6y = \tfrac{3}{2} \end{cases}$
6. $\begin{cases} y = 2x \\ y = x + 2 \end{cases}$

Reteach
Chapter 3

What you should learn :

| **3.2** | How to solve a linear system using algebraic methods. |

Correlation to Pupil's Textbook:

Mid-Chapter Test (p. 144)   Chapter Test (p. 169)
Exercises 7–12   Exercises 1–3

Examples   *Using Algebraic Methods to Solve Linear Systems*

a. Use the substitution method to solve the system. $\begin{cases} 6x + y = -2 & \textit{Equation 1} \\ 4x - 3y = 17 & \textit{Equation 2} \end{cases}$

$y = -6x - 2$	*Write Equation 1 in slope-intercept form.*
$4x - 3y = 17$	*Equation 2*
$4x - 3(-6x - 2) = 17$	*Substitute $-6x - 2$ for y in Equation 2.*
$4x + 18x + 6 = 17$	*Distributive Property*
$22x = 11$	*Simplify.*
$x = \frac{1}{2}$	*Solve for x.*
$y = -6x - 2$	*Slope-intercept form of Equation 1*
$y = -6(\frac{1}{2}) - 2$	*Substitute $\frac{1}{2}$ for x.*
$y = -5$	*Solve for y. The solution is $(\frac{1}{2}, -5)$.*

b. Use the linear combination method to solve the system. $\begin{cases} 5x - 3y = 14 & \textit{Equation 1} \\ 3x - 2y = 6 & \textit{Equation 2} \end{cases}$

$15x - 9y = 42$	*To obtain coefficients for x that differ in sign, multiply Equation 1 by 3.*
$\underline{-15x + 10y = -30}$	*Multiply Equation 2 by -5.*
$y = 12$	*Add the equations. From the result, you know that $y = 12$.*
$3x - 2y = 6$	*Equation 2*
$3x - 2(12) = 6$	*Substitute 12 for y.*
$x = 10$	*Solve for x. The solution is $(10, 12)$.*

Guidelines: To use algebraic methods to solve a system of linear equations:
- If one of the equations has a variable with a coefficient of 1, then use the substitution method as outlined on page 130 of the textbook.
- Otherwise, use the linear combination method as outlined on page 131 of the textbook.

EXERCISES

In 1–6, solve the system using an algebraic method.

1. $\begin{cases} 2x - y = 6 \\ 2x + 2y = -9 \end{cases}$

2. $\begin{cases} -2x + 3y = 5 \\ 3x - 2y = 0 \end{cases}$

3. $\begin{cases} 2x + 3y = 7 \\ x - 2y = -7 \end{cases}$

4. $\begin{cases} 2x - 5y = -4 \\ 4x + 3y = 5 \end{cases}$

5. $\begin{cases} 11x + 6y = 1 \\ 3x + 2y = -3 \end{cases}$

6. $\begin{cases} 4x - 3y = 5 \\ -8x + 6y = 17 \end{cases}$

Reteach
Chapter 3

Name _____

What you should learn :

3.3	How to write and use linear systems in problem solving.

Correlation to Pupil's Textbook:

Mid-Chapter Test (p. 144)   **Chapter Test (p. 169)**
Exercises 14, 15   Exercise 13

Example	*Using Linear Systems in Problem Solving*

In a basketball game between Westfield High School and Fairmount High School, the total number of points scored was 119. Fairmount's score was 49 points less than twice that of Westfield. How many points did each team score?

Verbal Model

$$\boxed{\text{Westfield points}} + \boxed{\text{Fairmount points}} = 119$$

$$\boxed{\text{Fairmount points}} = 2 \cdot \boxed{\text{Westfield points}} - 49$$

Labels
Number of Westfield points = W (points)
Number of Fairmount points = F (points)

System
$$\begin{cases} W + F = 119 \\ \quad F = 2W - 49 \end{cases}$$

	Equation 1
	Equation 2
$W + (2W - 49) = 119$	*Substitute $2W - 49$ for F in Equation 1.*
$3W = 168$	*Simplify.*
$W = 56$	*Solve for W.*
$F = 2(56) - 49$	*Substitute 56 for W in Equation 2.*
$F = 63$	*Solve for F.*

Westfield scored 56 points and Fairmount scored 63 points.

Guidelines: To solve a word problem, use the following:
- Write a verbal model.
- Assign labels.
- Write an algebraic model.
- Solve the model.
- Answer the question.

EXERCISES

1. The perimeter of a room is 88 feet. The length of the room is 8 feet more than the width. Find the dimensions of the room.

2. A business paid $262.26 for a mailing of 668 letters. Some of the letters needed $0.29 postage and the remainder needed $0.52 postage. How many letters were mailed at each postage rate?

Name _____

What you should learn :

| **3.4** | How to graph a system of linear inequalities. |

Correlation to Pupil's Textbook:

Chapter Test (p. 169)

Exercises 4–6, 15

Examples | *Graphing a System of Linear Inequalities*

a. Sketch the graph of the system of linear inequalities. $\begin{cases} x - y > 1 \\ x < 2 \\ y \geq 0 \end{cases}$

Begin by sketching the line $x - y = 1$ with a dashed line. Shade the half-plane that satisfies $x - y > 1$. Next, sketch the line $x = 2$ with a dashed line. Shade the half-plane that satisfies $x < 2$.

Finally, sketch the line $y = 0$ with a solid line. Shade the half-plane that satisfies $y \geq 0$.

The graph of the system is the region that has been shaded by all three inequalities, as shown in the first graph at the right.

The vertices of the graph are $(1,\ 0)$, $(2,\ 0)$, and $(2,\ 1)$, as shown in the second graph at the right.

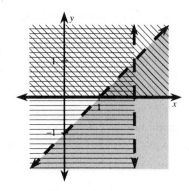 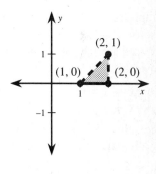

b. Sketch the graph of the system of linear inequalities. $\begin{cases} y \leq x \\ x - 2y \leq -1 \end{cases}$

Shade the half-plane that is the graph of $y \leq x$. Shade the half-plane that is the graph of $x - 2y \leq -1$. The graph of the system is the intersection of these two half-planes. You can obtain the vertex as follows.

$y = x$	*Equation 1*
$x - 2y = -1$	*Equation 2*
$x - 2(x) = -1$	*Substitute x for y into Equation 2.*
$x = 1$	*Solve for x.*
$y = 1$	*Use Equation 1 to solve for y.*

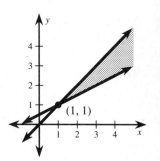

Guidelines: To graph a system of linear inequalities:
• Follow the guidelines on page 146 of the textbook.

EXERCISES

In 1–4, sketch the graph of the system. Find the vertices of the graph.

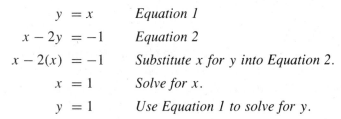

1. $\begin{cases} 3x + y \leq 5 \\ y \geq 1 \\ x \geq 1 \end{cases}$ **2.** $\begin{cases} x + 2y \leq 6 \\ x - y \geq 3 \\ x \geq 0 \end{cases}$ **3.** $\begin{cases} x + y > 2 \\ y \leq 2 \\ x \leq 2 \end{cases}$ **4.** $\begin{cases} x > 1 \\ x \leq 4 \\ y > 2 \\ y \leq 6 \end{cases}$

Reteach
Chapter 3

Name _____

3.5	How to solve a linear programming problem.

| **Example** | *Solving a Linear Programming Problem* |

Find the minimum and maximum values of the objective quantity
$C = 6x - 2y$, subject to the following constraints.

$$\begin{cases} x + y \le 9 \\ 4x + y \ge 12 \\ x \ge 0 \\ y \ge 0 \end{cases}$$

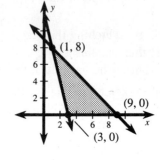

Graph the system of constraint inequalities using the methods
discussed in Lesson 3.4. The graph is shown at the right. Two of the
vertices are the intercepts (3, 0) and (9, 0). To find the third vertex,
solve the following system.

$$\begin{cases} x + y = 9 \\ 4x + y = 12 \end{cases}$$

The solution is (1, 8). To find the minimum and maximum values of
the objective function, evaluate $C = 6x - 2y$ at each of the three vertices.

$C = 6(3) - 2(0) = 18$ *Substitute (3, 0) for (x, y) in the objective quantity.*

$C = 6(9) - 2(0) = 54$ *Substitute (9, 0) for (x, y) in the objective quantity.*

$C = 6(1) - 2(8) = -10$ *Substitute (1, 8) for (x, y) in the objective quantity.*

The minimum value of C is -10, which occurs when $x = 1$ and $y = 8$.
The maximum value of C is 54, which occurs when $x = 9$ and $y = 0$.

Guidelines: To solve a linear programming problem:
- Graph the system of constraint inequalities.
- Find the vertices of the graph.
- Evaluate the objective quantity at each vertex.
- The least of these values is the minimum value of the objective quantity
  and the greatest is the maximum value (provided the objective quantity
  has a minimum value and a maximum value).

EXERCISES

In 1–3, find the minimum and maximum values of the objective quantity.

1. Objective quantity:
$C = 3x + 2y$
Constraints:
$$\begin{cases} 3x + 4y \le 25 \\ 3x - y \le 5 \\ x \ge 0 \\ y \ge 0 \end{cases}$$

2. Objective quantity:
$C = 2x + 5y$
Constraints:
$$\begin{cases} 2x + y \le 12 \\ x - 4y \le -3 \\ y \le x \\ x \ge 0 \\ y \ge 0 \end{cases}$$

3. Objective quantity:
$C = x + 3y$
Constraints:
$$\begin{cases} x + y \le 5 \\ x \ge 1 \\ y \ge 2 \end{cases}$$

Name _____

What you should learn :

| **3.6** | How to solve a system of three linear equations in three variables. |

Correlation to Pupil's Textbook:

Chapter Test (p. 169)
Exercises 10–12, 14

Example *Solving a Linear System Using Elimination*

Solve the linear system.
$$\begin{cases} x - 6y + z = 1 & \text{Equation 1} \\ -x + 2y - 4z = 3 & \text{Equation 2} \\ 7x - 10y + 3z = -25 & \text{Equation 3} \end{cases}$$

First write an equivalent system in which x has been eliminated from the second and third equations.

$$\begin{cases} x - 6y + z = 1 & \text{Equation 1} \\ -4y - 3z = 4 & \text{Add Equation 1 to Equation 2.} \\ 32y - 4z = -32 & \text{Add } -7 \text{ times Equation 1 to Equation 3.} \end{cases}$$

Next, use the second equation to eliminate y from the third equation.

$$\begin{cases} x - 6y + z = 1 & \text{Equation 1} \\ -4y - 3z = 4 & \text{Revised Equation 2} \\ -28z = 0 & \text{Add 8 times the revised Equation 2 to Equation 3.} \end{cases}$$

The system is now in triangular form. Using the third equation, you can determine that $z = 0$. Substituting this value into the second equation produces $y = -1$. Substituting $z = 0$ and $y = -1$ into the first equation produces $x = -5$. The solution is

$$x = -5, \quad y = -1, \quad \text{and} \quad z = 0.$$

The solution can also be represented by the ordered triple $(-5, -1, 0)$.

Guidelines: To solve three linear equations in three variables:
- Write the linear system in triangular form using the row operations discussed on page 158 of the textbook.
- Solve the third equation for z.
- Substitute the value of z into the second equation to solve for y.
- Substitute the values of y and z into the first equation to solve for x.

EXERCISES

In 1–3, solve the linear system.

1. $\begin{cases} x + 2y + z = 1 \\ 5y - 2z = -16 \\ z = 3 \end{cases}$

2. $\begin{cases} x + 2y + z = 6 \\ 2x - y + 3z = -2 \\ x + y - 2z = 0 \end{cases}$

3. $\begin{cases} x - y + z = 5 \\ 3x + 2y - z = -2 \\ 2x + y + 3z = 10 \end{cases}$

Name _____

What you should learn :

| 4.1 | How to add and subtract matrices and to multiply a matrix by a scalar. |

Correlation to Pupil's Textbook:

Mid-Chapter Test (p. 203)   **Chapter Test (p. 227)**
Exercises 1–4, 13   Exercises 1, 2

Examples | *Adding, Subtracting, and Multiplying Matrices*

a. To add or subtract matrices, they must have the same order.

$$\begin{bmatrix} 2 & 3 & -5 \\ -1 & 0 & 4 \end{bmatrix} - \begin{bmatrix} 0 & 1 & 3 \\ 3 & -2 & -1 \end{bmatrix}$$

$$= \begin{bmatrix} 2-0 & 3-1 & -5-3 \\ -1-3 & 0-(-2) & 4-(-1) \end{bmatrix}$$ *To subtract matrices, subtract corresponding entries.*

$$= \begin{bmatrix} 2 & 2 & -8 \\ -4 & 2 & 5 \end{bmatrix}$$ *Simplify.*

b. To multiply a matrix by a scalar (a real number), multiply each entry of the matrix by the scalar.

$$3\begin{bmatrix} -1 & 2 \\ 15 & 3 \end{bmatrix} = \begin{bmatrix} 3(-1) & 3(2) \\ 3(15) & 3(3) \end{bmatrix}$$ *Multiply each entry in the matrix by 3.*

$$= \begin{bmatrix} -3 & 6 \\ 45 & 9 \end{bmatrix}$$ *Simplify.*

c. To perform a combination of operations, do the scalar multiplication first. Then add or subtract matrices.

$$2\begin{bmatrix} 4 & 7 \\ 2 & 0 \end{bmatrix} - 3\begin{bmatrix} 1 & 2 \\ -1 & 5 \end{bmatrix} = \begin{bmatrix} 8 & 14 \\ 4 & 0 \end{bmatrix} - \begin{bmatrix} 3 & 6 \\ -3 & 15 \end{bmatrix}$$ *Multiply by scalar first.*

$$= \begin{bmatrix} 5 & 8 \\ 7 & -15 \end{bmatrix}$$ *Then subtract.*

Guidelines:
- To add or subtract matrices, they must have the same order.
- To add or subtract matrices, add or subtract corresponding entries.
- To multiply a matrix by a scalar, multiply each entry by the scalar.

EXERCISES

In 1–6, perform the indicated operation.

1. $\begin{bmatrix} 2 & -4 \\ -1 & 3 \end{bmatrix} + \begin{bmatrix} -3 & 2 \\ -2 & -3 \end{bmatrix}$

2. $\begin{bmatrix} 5 & 0 \\ -2 & 1 \\ 4 & -3 \end{bmatrix} - \begin{bmatrix} 4 & -6 \\ 2 & -2 \\ -1 & 3 \end{bmatrix}$

3. $-3\begin{bmatrix} 1 & 0 & -1 \\ 3 & \frac{1}{3} & -3 \\ -1 & -\frac{1}{3} & 0 \end{bmatrix}$

4. $2\begin{bmatrix} 2 & 10 & 15 \\ -1 & 3 & -6 \end{bmatrix}$

5. $\begin{bmatrix} 2 & -1 \\ 3 & 1 \end{bmatrix} - 2\begin{bmatrix} 4 & 0 \\ -1 & -1 \end{bmatrix}$

6. $3\begin{bmatrix} 3 & 6 & -1 \\ 0 & 5 & 2 \end{bmatrix} - 2\begin{bmatrix} 1 & 0 & 5 \\ -1 & 2 & 7 \end{bmatrix}$

Name _____

What you should learn :

4.2 How to multiply two matrices.

Correlation to Pupil's Textbook:

Mid-Chapter Test (p. 203)   **Chapter Test (p. 227)**
Exercises 5, 6, 11, 14   Exercises 3, 4, 19

| **Example** | *Multiplying Two Matrices* |

$$\begin{bmatrix} 2 & -1 & 0 \\ 3 & 4 & 1 \end{bmatrix} \begin{bmatrix} 0 & 1 \\ 4 & 3 \\ 5 & -1 \end{bmatrix}$$

$$= \begin{bmatrix} a & b \\ c & d \end{bmatrix}$$

To multiply two matrices, the number of columns of the left matrix must equal the number of rows of the right matrix. In this case, the product of a 2×3 and a 3×2 matrix is a 2×2 matrix.

$a = (2)(0) + (-1)(4) + (0)(5)$

$\quad = 0 - 4 + 0$

$\quad = -4$

To find entry a in the product matrix, multiply corresponding entries in the first row of the left matrix and the first column of the right matrix. Then add.

$b = (2)(1) + (-1)(3) + (0)(-1)$

$\quad = 2 - 3 + 0$

$\quad = -1$

To find entry b in the product matrix, multiply corresponding entries in the first row of the left matrix and the second column of the right matrix. Then add.

$c = (3)(0) + (4)(4) + (1)(5)$

$\quad = 0 + 16 + 5$

$\quad = 21$

To find entry c in the product matrix, multiply corresponding entries in the second row of the left matrix and the first column of the right matrix. Then add.

$d = (3)(1) + (4)(3) + (1)(-1)$

$\quad = 3 + 12 - 1$

$\quad = 14$

To find entry d in the product matrix, multiply corresponding entries in the second row of the left matrix and the second column of the right matrix. Then add.

The resulting product is $\begin{bmatrix} -4 & -1 \\ 21 & 14 \end{bmatrix}$.

Guidelines: To multiply two matrices:
- The number of columns of the left matrix must equal the number of rows of the right matrix.
- The product matrix will have the number of rows of the left matrix and the number of colmuns of the right matrix. (If A is an $m \times n$ matrix and B is an $n \times p$ matrix, then the product AB is an $m \times p$ matrix.)

EXERCISES

In 1–6, find the product.

1. $\begin{bmatrix} 0 & 1 \\ 4 & 3 \\ 5 & -1 \end{bmatrix} \begin{bmatrix} 2 & -1 & 0 \\ 3 & 4 & 1 \end{bmatrix}$

2. $\begin{bmatrix} 2 & 1 \\ 3 & -2 \end{bmatrix} \begin{bmatrix} -1 & 5 \\ 6 & 2 \end{bmatrix}$

3. $\begin{bmatrix} 1 & 0 & -1 \\ 2 & 3 & 3 \end{bmatrix} \begin{bmatrix} 2 & -1 & 0 \\ 0 & 5 & 3 \\ 1 & -2 & -1 \end{bmatrix}$

4. $\begin{bmatrix} 2 & 3 & 4 \end{bmatrix} \begin{bmatrix} -1 & 4 \\ 0 & 1 \\ 5 & 2 \end{bmatrix}$

5. $\begin{bmatrix} 2 & 4 \end{bmatrix} \begin{bmatrix} 5 \\ -3 \end{bmatrix}$

6. $\begin{bmatrix} 1 & -1 & 2 \\ 0 & 5 & 1 \\ -2 & 0 & -1 \end{bmatrix} \begin{bmatrix} -1 & 2 & 0 \\ 5 & -7 & 1 \\ 2 & 3 & -2 \end{bmatrix}$

Name _____

What you should learn :

4.3	How to evaluate the determinant of a 2×2 matrix or a 3×3 matrix.

Examples | *Evaluating Determinants*

a. The value of the determinant of a 2×2 matrix $\begin{vmatrix} a & b \\ c & d \end{vmatrix}$ is $ad - cb$.

$$\begin{vmatrix} 4 & -2 \\ 3 & 1 \end{vmatrix} = (4)(1) - (3)(-2) = 4 + 6 = 10$$

b. $\begin{vmatrix} 2 & 3 & -1 \\ 0 & 5 & 0 \\ -1 & 1 & 2 \end{vmatrix}$ *Determinant of a 3×3 matrix*

$= 2\begin{vmatrix} 5 & 0 \\ 1 & 2 \end{vmatrix} - 0\begin{vmatrix} 3 & -1 \\ 1 & 2 \end{vmatrix} + (-1)\begin{vmatrix} 3 & -1 \\ 5 & 0 \end{vmatrix}$ *Expand by minors along the first column.*

$= 2(10 - 0) - 0 - 1(0 + 5)$ *Evaluate the 2×2 determinants.*

$= 20 - 5 = 15$ *Simplify.*

c. $\begin{vmatrix} 2 & 3 & -1 \\ 0 & 5 & 0 \\ -1 & 1 & 2 \end{vmatrix}$ *Determinant of a 3×3 matrix*

$= \begin{vmatrix} 2 & 3 & -1 \\ 0 & 5 & 0 \\ -1 & 1 & 2 \end{vmatrix} \begin{matrix} 2 & 3 \\ 0 & 5 \\ -1 & 1 \end{matrix}$ *Write a 4th column by copying the 1st column.*
     Write a 5th column by copying the 2nd column.

$= (2)(5)(2) + (3)(0)(-1) + (-1)(0)(1)$ *Add products of "top-to-bottom" diagonals.*

$\quad - (-1)(5)(-1) - (1)(0)(2) - (2)(0)(3)$ *Subtract products of "bottom-to-top" diagonals.*

$= 20 + 0 + 0 - 5 - 0 - 0 = 15$ *Simplify.*

Guidelines: To evaluate a determinant:
- The determinant of a 2×2 matrix is $\begin{vmatrix} a & b \\ c & d \end{vmatrix} = ad - cb$.
- One way to find the determinant of a 3×3 matrix is expansion by minors as outlined on page 188 of the textbook.
- Another way to find the determinant of a 3×3 matrix is the diagonal method as shown on page 189 of the textbook.

EXERCISES

In 1–6, evaluate the determinant of the matrix.

1. $\begin{bmatrix} 7 & -1 \\ 6 & -2 \end{bmatrix}$ **2.** $\begin{bmatrix} 6 & 4 \\ -2 & 1 \end{bmatrix}$ **3.** $\begin{bmatrix} 3 & -1 & 6 \\ 2 & 0 & 4 \\ 1 & 6 & 2 \end{bmatrix}$

4. $\begin{bmatrix} 0 & 2 & 3 \\ 1 & -1 & 4 \\ 3 & 0 & 2 \end{bmatrix}$ **5.** $\begin{bmatrix} 0 & -1 & 2 \\ 3 & 5 & 0 \\ 1 & -1 & 3 \end{bmatrix}$ **6.** $\begin{bmatrix} 3 & 0 & 1 \\ -1 & 4 & -1 \\ 5 & -2 & 0 \end{bmatrix}$

Scrambled answers for first row of exercises: 14, 0, −8

Name _____

What you should learn :

| **4.4** | How to find and use the inverse of a 2×2 matrix. |

Correlation to Pupil's Textbook:

Mid-Chapter Test (p. 203)   Chapter Test (p. 227)
Exercises 9–10   Exercises 5–7, 17

Examples *Finding and Using the Inverse of a 2 × 2 Matrix*

a. Find the inverse of $A = \begin{bmatrix} 3 & 1 \\ 2 & 2 \end{bmatrix}$.

$|A| = (3)(2) - (2)(1) = 4$ *Evaluate the determinant of A.*

$A^{-1} = \frac{1}{4}\begin{bmatrix} 2 & -1 \\ -2 & 3 \end{bmatrix} = \begin{bmatrix} \frac{1}{2} & -\frac{1}{4} \\ -\frac{1}{2} & \frac{3}{4} \end{bmatrix}$ *The inverse of $A = \begin{bmatrix} a & b \\ c & d \end{bmatrix}$ is $\frac{1}{|A|}\begin{bmatrix} d & -b \\ -c & a \end{bmatrix}$.*

b. Solve the matrix equation for X.

$\underbrace{\begin{bmatrix} 3 & 2 \\ 4 & 3 \end{bmatrix}}_{A} X = \underbrace{\begin{bmatrix} 12 & -11 & 2 \\ 17 & -16 & 3 \end{bmatrix}}_{B}$

$A^{-1} = \frac{1}{9-8}\begin{bmatrix} 3 & -2 \\ -4 & 3 \end{bmatrix}$ *Calculate the inverse of A.*

$= \begin{bmatrix} 3 & -2 \\ -4 & 3 \end{bmatrix}$

$\begin{bmatrix} 3 & -2 \\ -4 & 3 \end{bmatrix}\begin{bmatrix} 3 & 2 \\ 4 & 3 \end{bmatrix} X = \begin{bmatrix} 3 & -2 \\ -4 & 3 \end{bmatrix}\begin{bmatrix} 12 & -11 & 2 \\ 17 & -16 & 3 \end{bmatrix}$ *To solve the equation for X, multiply both sides of the equation by A^{-1} on the left.*

$X = \begin{bmatrix} 36-34 & -33+32 & 6-6 \\ -48+51 & 44-48 & -8+9 \end{bmatrix}$ *The left side of the equation is X because $A^{-1}AX = IX = X$.*

$X = \begin{bmatrix} 2 & -1 & 0 \\ 3 & -4 & 1 \end{bmatrix}$ *Simplify.*

Guidelines: To solve a matrix equation $AX = B$:
- Calculate A^{-1}.
- Multiply both sides of the equation by A^{-1} *on the left.*
- X is the product $A^{-1}B$.

EXERCISES

In 1–3, find the inverse of the matrix.

1. $\begin{bmatrix} -3 & -7 \\ 2 & 5 \end{bmatrix}$ **2.** $\begin{bmatrix} 2 & 5 \\ 4 & 11 \end{bmatrix}$ **3.** $\begin{bmatrix} 2 & 1 \\ 1 & 2 \end{bmatrix}$

In 4 and 5, solve the matrix equation.

4. $\begin{bmatrix} 2 & 2 \\ 1 & 3 \end{bmatrix} X = \begin{bmatrix} 12 & 2 \\ 16 & -5 \end{bmatrix}$ **5.** $\begin{bmatrix} 4 & -6 \\ 3 & -4 \end{bmatrix} X = \begin{bmatrix} 6 & 0 & -2 \\ 5 & 1 & -1 \end{bmatrix}$

Name _____

4.5	How to use matrices to solve a system of linear equations.

Correlation to Pupil's Textbook:

Chapter Test (p. 227)
Exercises 11, 14

Examples	*Solving Systems Using Inverse Matrices*

a. Solve the system. $\begin{cases} 5x + 2y = 3 \\ 4x + 2y = 4 \end{cases}$

$\underbrace{\begin{bmatrix} 5 & 2 \\ 4 & 2 \end{bmatrix}}_{A} \underbrace{\begin{bmatrix} x \\ y \end{bmatrix}}_{X} = \underbrace{\begin{bmatrix} 3 \\ 4 \end{bmatrix}}_{B}$ *Write the matrix equation for the system.*

$A^{-1} = \dfrac{1}{10 - 8}\begin{bmatrix} 2 & -2 \\ -4 & 5 \end{bmatrix} = \begin{bmatrix} 1 & -1 \\ -2 & \frac{5}{2} \end{bmatrix}$ *Find the inverse of the coefficient matrix A.*

$\begin{bmatrix} x \\ y \end{bmatrix} = \begin{bmatrix} 1 & -1 \\ -2 & \frac{5}{2} \end{bmatrix}\begin{bmatrix} 3 \\ 4 \end{bmatrix} = \begin{bmatrix} -1 \\ 4 \end{bmatrix}$ *To find X, multiply B by A^{-1} on the left.*

The solution is $x = -1$ and $y = 4$.

b. Solve the system of linear equations, given the inverse of the coefficient matrix.

$\begin{cases} 3x - 5y + z = 5 \\ -4x + 7y - z = -3 \\ 8x - 13y + 2z = 4 \end{cases}$ $A^{-1} = \begin{bmatrix} -1 & 3 & 2 \\ 0 & 2 & 1 \\ 4 & 1 & -1 \end{bmatrix}$

To find $\begin{bmatrix} x \\ y \\ z \end{bmatrix}$, multiply the column of constants by A^{-1} on the left.

$\begin{bmatrix} x \\ y \\ z \end{bmatrix} = \begin{bmatrix} -1 & 3 & 2 \\ 0 & 2 & 1 \\ 4 & 1 & -1 \end{bmatrix}\begin{bmatrix} 5 \\ -3 \\ 4 \end{bmatrix} = \begin{bmatrix} -6 \\ -2 \\ 13 \end{bmatrix}$

The solution is $x = -6$, $y = -2$, and $z = 13$.

Guidelines: To solve a linear system using an inverse matrix:
- Write the corresponding matrix equation, $AX = B$.
- Find the inverse of the coefficient matrix, A^{-1}.
- Multiply the inverse of the coefficient matrix by the matrix of constants to obtain $X = A^{-1}B$.

EXERCISES

In 1–4, use an inverse matrix to solve the linear system.

1. $\begin{cases} 2x - y = 1 \\ -3x + 2y = 0 \end{cases}$

2. $\begin{cases} 3x + 4y = -4 \\ 4x + 5y = -7 \end{cases}$

3. $\begin{cases} 6x - 5y = 3 \\ 3x - 2y = 3 \end{cases}$

4. $\begin{cases} 2x - 11y + 2z = 15 \\ -4y + z = -20, \\ -x + 6y - z = -6 \end{cases}$ given that $A^{-1} = \begin{bmatrix} 2 & -1 & 3 \\ 1 & 0 & 2 \\ 4 & 1 & 8 \end{bmatrix}$

Name _____

What you should learn :

| **4.6** | How to solve a system of linear equations using an augmented matrix. |

Correlation to Pupil's Textbook:

Chapter Test (p. 227)
Exercises 12, 15

Example *Solving a System Using an Augmented Matrix*

$$\begin{cases} 2x - y + 12z = -16 \\ x + 3y - 8z = 13 \\ -5x + y + 2z = 44 \end{cases}$$ *System of linear equations*

$$\begin{bmatrix} 2 & -1 & 12 & \vdots & -16 \\ 1 & 3 & -8 & \vdots & 13 \\ -5 & 1 & 2 & \vdots & 44 \end{bmatrix}$$ *Write the augmented matrix for the system.*

$\begin{matrix} R_2 \\ R_1 \end{matrix}$ $\begin{bmatrix} 1 & 3 & -8 & \vdots & 13 \\ 2 & -1 & 12 & \vdots & -16 \\ -5 & 1 & 2 & \vdots & 44 \end{bmatrix}$ *To obtain a 1 in the top-left corner, interchange Rows 1 and 2.*

$-2R_1 + R_2 \rightarrow$ $\begin{bmatrix} 1 & 3 & -8 & \vdots & 13 \\ 0 & -7 & 28 & \vdots & -42 \\ -5 & 1 & 2 & \vdots & 44 \end{bmatrix}$ *Add −2 times the first row to the second row.*

$5R_1 + R_3 \rightarrow$ $\begin{bmatrix} 1 & 3 & -8 & \vdots & 13 \\ 0 & -7 & 28 & \vdots & -42 \\ 0 & 16 & -38 & \vdots & 109 \end{bmatrix}$ *Add 5 times the first row to the third row.*

$-\frac{1}{7}R_2 \rightarrow$ $\begin{bmatrix} 1 & 3 & -8 & \vdots & 13 \\ 0 & 1 & -4 & \vdots & 6 \\ 0 & 16 & -38 & \vdots & 109 \end{bmatrix}$ *Multiply the second row by $-\frac{1}{7}$.*

$-16R_2 + R_3 \rightarrow$ $\begin{bmatrix} 1 & 3 & -8 & \vdots & 13 \\ 0 & 1 & -4 & \vdots & 6 \\ 0 & 0 & 26 & \vdots & 13 \end{bmatrix}$ *Add −16 times the second row to the third row.*

$\frac{1}{26}R_3 \rightarrow$ $\begin{bmatrix} 1 & 3 & -8 & \vdots & 13 \\ 0 & 1 & -4 & \vdots & 6 \\ 0 & 0 & 1 & \vdots & \frac{1}{2} \end{bmatrix}$ *Multiply the third row by $\frac{1}{26}$.*

$$\begin{cases} x + 3y - 8z = 13 \\ y - 4z = 6 \\ z = \frac{1}{2} \end{cases}$$ *Corresponding linear system*

Use substitution to find that the solution is $x = -7$, $y = 8$, and $z = \frac{1}{2}$.

Guidelines: To solve a system using an augmented matrix:
- Follow the example given above, using elementary row operations as listed on page 210 of the textbook.

EXERCISES

In 1 and 2, solve the system using an augmented matrix.

1. $\begin{cases} x - 2y + 3z = 9 \\ -x + 3y = -4 \\ 2x - 5y + 5z = 17 \end{cases}$

2. $\begin{cases} x + y - z = 1 \\ x - y - z = 2 \\ x + y + z = -3 \end{cases}$

Name _____

Correlation to Pupil's Textbook:

Chapter Test (p. 227)

Exercises 13, 16

4.7	How to solve a system of linear equations using Cramer's Rule.

Example   *Solving a System Using Cramer's Rule*

$$\begin{cases} 2x - 2y + 2z = -8 \\ 3x + y + z = 12 \\ 2x - y + z = -1 \end{cases}$$ *System of linear equations*

$$\begin{vmatrix} 2 & -2 & 2 \\ 3 & 1 & 1 \\ 2 & -1 & 1 \end{vmatrix} \begin{matrix} 2 & -2 \\ 3 & 1 \\ 2 & -1 \end{matrix} = 2 - 4 - 6 - 4 - (-2) - (-6)$$

First, calculate the determinant of the coefficient matrix. Because this number is not zero, it is the denominator of each variable.

$$= -4$$

$$x = \frac{\begin{vmatrix} -8 & -2 & 2 \\ 12 & 1 & 1 \\ -1 & -1 & 1 \end{vmatrix}}{-4} = \frac{-12}{-4} = 3$$

For the numerator of x, replace the first column of the coefficient matrix by the column of constants.

$$y = \frac{\begin{vmatrix} 2 & -8 & 2 \\ 3 & 12 & 1 \\ 2 & -1 & 1 \end{vmatrix}}{-4} = \frac{-20}{-4} = 5$$

For the numerator of y, replace the second colun of the coefficient matrix by the column of constan

$$z = \frac{\begin{vmatrix} 2 & -2 & -8 \\ 3 & 1 & 12 \\ 2 & -1 & -1 \end{vmatrix}}{-4} = \frac{8}{-4} = -2$$

For the numerator of z, replace the third column of the coefficient matrix by the column of constants.

The solution is $x = 3$, $y = 5$, and $z = -2$.

Guidelines: To solve a system using Cramer's Rule:
- Calculate the determinant of the coefficient matrix. This number must not be zero because it is the value of the denominator for each variable.
- To find the numerator of each variable, calculate the determinant of the matrix found by using the column of constants in place of the column of coefficients for that particular variable.
- The variable is equal to the numerator divided by the denominator.

EXERCISES

In 1–6, use Cramer's Rule to solve the linear system.

1. $\begin{cases} 2x + y = 1 \\ -x + y = 7 \end{cases}$

2. $\begin{cases} 3x + 4y = 2 \\ 2x + y = 3 \end{cases}$

3. $\begin{cases} 3x - 2y = 22 \\ x + 4y = -2 \end{cases}$

4. $\begin{cases} -x + 3y + 4z = -16 \\ 2x + y + 2z = 2 \\ 5x + y + 3z = 11 \end{cases}$

5. $\begin{cases} 2x - y - z = 4 \\ -x - y + 2z = -5 \\ x + 2y + z = 3 \end{cases}$

6. $\begin{cases} x + 2y + z = 5 \\ 2x - y - 3z = 5 \\ -2x + 3y + z = -11 \end{cases}$

Reteach
Chapter 5

Name _____

What you should learn :

5.1	How to solve a quadratic equation by finding square roots.

Correlation to Pupil's Textbook:

Mid-Chapter Test (p. 251)
Exercises 1–3, 15–17, 19, 20

Examples *Solving a Quadratic Equation by Finding Square Roots*

a. $12x^2 = 132$ *Original equation*

$\quad x^2 = 11$ *To isolate x^2, divide both sides by 12.*

$\quad x = \pm\sqrt{11}$ *Find square roots.*

b. $\frac{1}{2}x^2 + 9 = 41$ *Original equation*

$\quad \frac{1}{2}x^2 = 32$ *To isolate $\frac{1}{2}x^2$, subtract 9 from both sides.*

$\quad x^2 = 64$ *Multiply both sides by 2.*

$\quad x = \pm 8$ *Find square roots.*

c. The lengths of the two sides of a right triangle are 39 centimeters and 52 centimeters. Find the length of the hypotenuse of the right triangle.

$c^2 = a^2 + b^2$ *Pythagorean Theorem*

$c^2 = 39^2 + 52^2$ *Substitute 39 for a and 52 for b.*

$c^2 = 4225$ *Simplify.*

$c = 65$ *Find the positive square root.*

The length of the hypotenuse is 65 centimeters.

Guidelines: To solve the equation $ax^2 + b = c$ by finding square roots:
- Isolate the term with the variable on one side of the equation.
- Divide both sides by the coefficient a.
- Take the square root of both sides.

EXERCISES

In 1–9, solve the equation.

1. $3x^2 = 192$ **2.** $5x^2 = 65$ **3.** $\frac{1}{4}x^2 = 8$

4. $\frac{2}{5}x^2 = 10$ **5.** $x^2 - 45 = 0$ **6.** $c^2 = 9 + 16$

7. $25 + b^2 = 169$ **8.** $\frac{3}{4}x^2 + 5 = 32$ **9.** $-\frac{1}{2}x^2 + 20 = 2$

10. The lengths of the sides of a right triangle are 6 inches and 7 inches. Find the length of the hypotenuse.

11. An object is dropped from the top of a 550 foot building. How long does it take for the object to hit the ground? Use the formula $h = -16t^2 + s$ and round the result to two decimal places.

Scrambled answers for Exercises 1, 4, and 7: ± 8, ± 12, ± 5

© D.C. Heath and Company

Algebra 2   *Chapter 5 ▪ Quadratic Equations and Parabolas* **29**

Reteach
Chapter 5

Name _____

What you should learn :

| 5.2 | How to graph a quadratic equation. |

Correlation to Pupil's Textbook:

Mid-Chapter Test (p. 251)   Chapter Test (p. 281)
Exercises 7–12, 18   Exercises 13, 14, 19

Examples   *Sketching the Graph of a Quadratic Equation*

a. Sketch the graph of $y = \frac{1}{4}x^2 + x$.

To find the x-coordinate of the vertex, substitute $\frac{1}{4}$ for a and 1 for b in the formula $-\frac{b}{2a}$.

$$-\frac{b}{2a} = -\frac{1}{2(\frac{1}{4})} = -\frac{1}{\frac{1}{2}} = -2$$

Because a is positive, the parabola opens up. Construct a table of values, choosing x-values to the left and right of the vertex.

x	−6	−4	−2	0	2
y	3	0	−1	0	3

Plot the points and connect them with a $\bigcup$-shaped graph, as shown at the right.

b. Sketch the graph of $y = -2x^2 + 4x + 1$.

To find the x-coordinate of the vertex, substitute -2 for a and 4 for b in the formula $-\frac{b}{2a}$.

$$-\frac{b}{2a} = -\frac{4}{2(-2)} = -\frac{4}{-4} = 1$$

Because a is negative, the parabola opens down. Construct a table of values, choosing x-values to the left and right of the vertex.

x	−1	0	1	2	3
y	−5	1	3	1	−5

Plot the points and connect them with a $\bigcup$-shaped graph, as shown at the right.

Guidelines: To sketch the graph of $y = ax^2 + bx + c$:
- Decide whether the parabola opens up ($a > 0$) or down ($a < 0$).
- Find the x-coordinate of the vertex by evaluating $-\frac{b}{2a}$.
- Construct a table of values, choosing x-values to the left and right of the vertex.
- Plot the points given in the table.
- Draw a $\bigcup$-shaped graph through the points.

EXERCISES

In 1–6, sketch the graph of the equation.

1. $y = x^2 + 2x + 1$ **2.** $y = x^2 - 4x + 3$ **3.** $y = -x^2 + 4$

4. $y = -2x^2 - 8x - 5$ **5.** $y = x^2 + 6x$ **6.** $y = \frac{1}{2}x^2 + 3x + 2$

30   *Chapter 5 ▪ Quadratic Equations and Parabolas*   *Algebra 2*

© D.C. Heath and Company

Name _____

What you should learn :

5.3	How to solve a quadratic equation by completing the square.

Correlation to Pupil's Textbook:

Mid-Chapter Test (p. 251) **Chapter Test (p. 281)**
Exercises 4–6, 13, 14   Exercises 1, 2, 20

Examples	*Solving a Quadratic Equation by Completing the Square*

a. $x^2 - 2x - 2 = 0$ *Original equation*

$\quad x^2 - 2x = 2$ *To isolate the terms containing x, add 2 to both sides.*

$\quad \left(\frac{-2}{2}\right)^2 = 1$ *Take half of -2 (the coefficient of x) and square it.*

$\quad x^2 - 2x + 1 = 2 + 1$ *To complete the square on the left, add 1 to both sides.*

$\quad (x-1)^2 = 3$ *Write the left side as the square of a binomial.*

$\quad x - 1 = \pm\sqrt{3}$ *Take square root of both sides.*

$\quad x = 1 \pm \sqrt{3}$ *To solve for x, add 1 to both sides.*

The solutions are $1 + \sqrt{3}$ and $1 - \sqrt{3}$.

b. $4x^2 - 6x + 1 = 0$ *Original equation*

$\quad 4x^2 - 6x = -1$ *To isolate the terms containing x, subtract 1 from both sides.*

$\quad x^2 - \frac{3}{2}x = -\frac{1}{4}$ *To make the coefficient of the x^2-term 1, divide both sides by 4.*

$\quad \left[\frac{1}{2}\left(-\frac{3}{2}\right)\right]^2 = \frac{9}{16}$ *Take half of $-\frac{3}{2}$ (the coefficient of x) and square it.*

$\quad x^2 - \frac{3}{2}x + \frac{9}{16} = -\frac{1}{4} + \frac{9}{16}$ *To complete the square on the left, add $\frac{9}{16}$ to both sides.*

$\quad \left(x - \frac{3}{4}\right)^2 = \frac{5}{16}$ *Write the left side as the square of a binomial.*

$\quad x - \frac{3}{4} = \pm\frac{\sqrt{5}}{4}$ *Take square root of both sides.*

$\quad x = \frac{3}{4} \pm \frac{\sqrt{5}}{4}$ *To solve for x, add $\frac{3}{4}$ to both sides.*

The solutions are $\frac{3}{4} + \frac{\sqrt{5}}{4}$ and $\frac{3}{4} - \frac{\sqrt{5}}{4}$.

Guidelines: To solve a quadratic equation by completing the square:
- Isolate the terms containing x on the left side of the equation.
- Divide both sides by the coefficient of x^2.
- Take half of the coefficient of x and square it.
- Add this number to both sides of the equation.
- Write the left side as the square of a binomial.
- Take the square root of both sides. Then solve for x.

EXERCISES

In 1–6, solve the equation by completing the square.

1. $x^2 + 6x - 7 = 0$ **2.** $x^2 - 10x + 11 = 0$ **3.** $x^2 - x - 1 = 0$

4. $4x^2 + 20x + 9 = 0$ **5.** $4x^2 - 16x + 13 = 0$ **6.** $2x^2 - x - 2 = 0$

What you should learn :

| 5.4 | How to solve a quadratic equation using the quadratic formula. |

Examples   *Solving a Quadratic Equation by the Quadratic Formula*

a. $x^2 - 2x + 1 = 3x + 5$ *Original equation*

     $x^2 - 5x - 4 = 0$ *Write in standard form: $a = 1$, $b = -5$, and $c = -4$.*

     $x = \dfrac{-(-5) \pm \sqrt{(-5)^2 - 4(1)(-4)}}{2(1)}$ *Quadratic formula*

     $x = \dfrac{5 \pm \sqrt{41}}{2}$ *Simplify.*

The solutions are $x = \frac{5+\sqrt{41}}{2}$ and $x = \frac{5-\sqrt{41}}{2}$.

b. $8x^2 + 28x = 12$ *Original equation*

     $8x^2 + 28x - 12 = 0$ *Write the equation in standard form.*

     $2x^2 + 7x - 3 = 0$ *Divide by the common multiple 4.*

     $x = \dfrac{-7 \pm \sqrt{7^2 - 4(2)(-3)}}{2(2)}$ *Substitute 2 for a, 7 for b, and -3 for c into quadratic formula.*

     $x = \dfrac{-7 \pm \sqrt{73}}{4}$ *Simplify.*

The solutions are $x = \frac{-7+\sqrt{73}}{4}$ and $x = \frac{-7-\sqrt{73}}{4}$.

c. The discriminant of $3x^2 - x + 4 = 0$ is $b^2 - 4ac = (-1)^2 - 4(3)(4) = -47$.
Because the discriminant is negative the equation has no real solution.

Guidelines: To solve a quadratic equation by the quadratic formula:
- Write the equation in standard form: $ax^2 + bx + c = 0$.
- Divide by the common multiple, if there is one.
- Identify a, b, and c. Then evaluate the quadratic formula: $x = \dfrac{-b \pm \sqrt{b^2 - 4ac}}{2a}$.

EXERCISES

In 1–9, use the quadratic formula to solve the equation.

1. $3x^2 + 15x + 6 = 0$ **2.** $2x^2 - 3x - 2 = 0$ **3.** $7x^2 - 5x = 1$

4. $6x = 6x^2 - 2$ **5.** $4x^2 = -7x + 3$ **6.** $10x^2 = 10x + 30$

7. $14x^2 - 36x + 20 = 2x^2 - 7$ **8.** $x^2 + 4x + 2 = 0$ **9.** $4x^2 + 12x - 135 = 0$

Scrambled answers for first column of exercises: $\frac{1}{2} \pm \frac{\sqrt{21}}{6}$, $-\frac{5}{2} \pm \frac{\sqrt{17}}{2}$, $\frac{3}{2}$

Name _____

What you should learn :

5.5	How to add, subtract, and multiply complex numbers.

Examples	*Performing Algebraic Operations with Complex Numbers*

a. $-\sqrt{-36} = -i\sqrt{36} = -6i$

b. $i^5 = i^2 \cdot i^2 \cdot i = (-1)(-1)i = i$

c. $(i\sqrt{2})^2 = i^2(\sqrt{2})^2 = (-1)(2) = -2$

d. $(2+3i) - (5-6i)$
$\qquad = 2+3i-5+6i$ *Distributive Property*
$\qquad = 2-5+3i+6i$ *Collect real and imaginary parts.*
$\qquad = -3+9i$ *Simplify.*

e. $-2i(4-3i)$
$\qquad = -8i + 6i^2$ *Distributive Property*
$\qquad = -8i + 6(-1)$ *Substitute -1 for i^2.*
$\qquad = -6 - 8i$ *Write the number in standard form.*

f. $(4-i)(3-2i)$
$\qquad = 4(3-2i) - i(3-2i)$ *Distributive Property*
$\qquad = 12 - 8i - 3i + 2i^2$ *Distributive Property*
$\qquad = 12 - 8i - 3i - 2$ *Substitute -1 for i^2.*
$\qquad = 10 - 11i$ *Write the number in standard form.*

Guidelines: To perform algebraic operations with complex numbers:
- If a is a positive real number, then $\sqrt{-a} = i\sqrt{a}$.
- The standard form for a complex number is $a + bi$.
- The Associative Properties of addition and multiplication, the Commutative Properties of addition and multiplication, and the Distributive Property hold for complex numbers.

EXERCISES

In 1–3, simplify.

1. $-\sqrt{-16}$ **2.** i^{15} **3.** $-i\sqrt{-2}$

In 4–12, perform the indicated operations.

4. $(4+3i) - (2+i)$ **5.** $(4+3i)+(-2-5i)-(7-i)$ **6.** $2(4+i) - i(4+i)$

7. $(-i\sqrt{3})^2$ **8.** $(7+6i)(7-6i)$ **9.** $(5-3i)(1+2i)$

10. $i(1+3i)(3-7i)$ **11.** $(10+3i)(7+2i)$ **12.** $(2+5i)^2$

Scrambled answers for first column of exercises: $-2+24i$, $-4i$, -3, $2+2i$

Name _____

What you should learn :

| 5.6 | How to solve a quadratic equation that has complex-number solutions. |

Correlation to Pupil's Textbook:

Chapter Test (p. 281)
Exercises 4, 22–24

Examples | *Solving Quadratic Equations*

a. $x^2 - 6x + 10 = 0$　　　　　　*Original equation in standard form*

$$x = \frac{6 \pm \sqrt{(-6)^2 - 4(1)(10)}}{2(1)}$$　*Substitute 1 for a, −6 for b, and 10 for c in the quadratic formula.*

$$x = \frac{6 \pm \sqrt{-4}}{2}$$　　　*Simplify.*

$$x = \frac{6 \pm 2i}{2}$$　　　　*Write with imaginary unit i.*

$$x = 3 \pm i$$　　　　*Write the numbers in standard form.*

The solutions are $3 - i$ and $3 + i$.

b. 　　$3x^2 + 5 = 3x$　　　　　*Original equation*

$3x^2 - 3x + 5 = 0$　　　*Write the equation in standard form.*

$$x = \frac{3 \pm \sqrt{(-3)^2 - 4(3)(5)}}{2(3)}$$　*Substitute 3 for a, −3 for b, and 5 for c in the quadratic formula.*

$$x = \frac{3 \pm \sqrt{-51}}{6}$$　　　*Simplify.*

$$x = \frac{3 \pm i\sqrt{51}}{6}$$　　　*Write with imaginary unit i.*

$$x = \tfrac{1}{2} \pm \tfrac{\sqrt{51}}{6} i$$　　　*Write the numbers in standard form.*

The solutions are $\tfrac{1}{2} - \tfrac{\sqrt{51}}{6} i$ and $\tfrac{1}{2} + \tfrac{\sqrt{51}}{6} i$.

Guidelines:　To solve a quadratic equation that has complex-number solutions:
- Use the quadratic formula.
- Write the complex-number solutions in standard form.

EXERCISES

In 1–9, solve the equation.

1. $x^2 + 5x + 7 = 0$　　　　**2.** $x^2 + 2x + 4 = 0$　　　　**3.** $x^2 - 4x + 13 = 0$

4. $x^2 - 2x + 3 = 0$　　　　**5.** $2x^2 - 2x + 3 = 0$　　　　**6.** $x^2 = -x - 1$

7. $12x^2 + 2 = 9x^2 - x$　　　**8.** $5x^2 + 2x - 2 = 6x^2 - x + 3$　　　**9.** $7x^2 - 3x + 15 = 0$

Scrambled answers for first column of exercises: $1 \pm i\sqrt{2},\ -\tfrac{1}{6} \pm \tfrac{\sqrt{23}}{6}i,\ -\tfrac{5}{2} \pm \tfrac{\sqrt{3}}{2}i$

Reteach
Chapter 5

Name _____

What you should learn :

5.7	How to sketch the graph of a quadratic inequality.

Correlation to Pupil's Textbook:

Chapter Test (p. 281)
Exercises 15, 16

Examples   *Sketching the Graph of a Quadratic Inequality*

a. Sketch the graph of the inequality $y \geq -2x^2 - 4x + 1$.

Begin by sketching the graph of $y = -2x^2 - 4x + 1$. The vertex is $(-1,\ 3)$.

Test $(0,\ 0)$, inside the parabola: $0 \overset{?}{\geq} -2(0) - 4(0) + 1$

Test $(0,\ 4)$, outside the parabola: $4 \overset{?}{\geq} -2(0) - 4(0) + 1$

Because 4 is greater than 1, the ordered pair $(0,\ 4)$ is a solution. Thus, the graph is the region on and outside the parabola, as shown at the right.

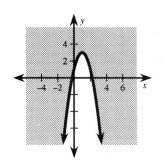

b. Sketch the intersection of the graphs of $y > x^2 - 4x + 4$ and $y < \frac{1}{4}x^2 - x + 3$.

The dashed parabola $y = x^2 - 4x + 4$ opens up and has its vertex at $(2,\ 0)$. The graph of the inequality $y > x^2 - 4x + 4$ is the region inside the parabola.

The dashed parabola $y = \frac{1}{4}x^2 - x + 3$ opens up and has its vertex at $(2,\ 2)$. The graph of the inequality $y < \frac{1}{4}x^2 - x + 3$ is the region outside the parabola.

The intersection of the graphs is the region inside the parabola $y = x^2 - 4x + 4$ and outside the parabola $y = \frac{1}{4}x^2 - x + 3$, as shown at the right.

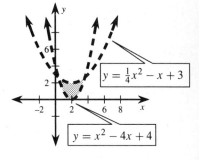

Guidelines: To sketch the graph of a quadratic inequality:
- Follow the steps outlined on page 271 of the textbook.

EXERCISES

In 1–6, sketch the graph of the inequality.

1. $y < x^2 + 2$ **2.** $y \leq -x^2 + 2x$ **3.** $y \geq 2x^2 + 4x + 3$

4. $y \geq \frac{1}{4}x^2 - 3x + 6$ **5.** $y \geq -3x^2 - 3x + 1$ **6.** $y < -x^2 - 4x - 4$

In 7–9, sketch the intersection of the graphs of the inequalities.

7. $y \geq x^2$ **8.** $y \leq -2x^2 + 4$ **9.** $y < \frac{1}{2}x^2 + 2x - 2$

     $y \leq 4x - x^2$ $y \geq x^2 - 2x + 2$ $y > x^2 - 2$

Reteach
Chapter 6

What you should learn :

6.1	How to identify a relation and a function and how to evaluate a function.

Correlation to Pupil's Textbook:

Mid-Chapter Test (p. 305)	Chapter Test (p. 339)
Exercises 1–6, 19	Exercises 1, 17

Examples — *Identifying and Evaluating Functions*

a. The following relation *is* a function of x because each x-value corresponds to exactly one y-value.

$$\{(0,\ 6),\ (1,\ 2),\ (2,\ 3),\ (3,\ -1),\ (4,\ 5),\ (5,\ 2)\}$$

The following relation is *not* a function of x because the x-value 3 corresponds to two values of y: -1 and 6.

$$\{(0,\ 4),\ (1,\ 3),\ (2,\ 2),\ (3,\ -1),\ (3,\ 6),\ (4,\ 5)\}$$

b. Find $f(-2)$ for $f(x) = -x^2 + 2x + 4$.

$$\begin{aligned} f(-2) &= -(-2)^2 + 2(-2) + 4 & \textit{Substitute } -2 \textit{ for } x. \\ &= -4 - 4 + 4 & \textit{Simplify.} \\ &= -4 & \textit{Add and subtract terms.} \end{aligned}$$

c. Find $f(\frac{1}{2})$ for $f(x) = 2x^2 + x - 1$.

$$\begin{aligned} f(\tfrac{1}{2}) &= 2(\tfrac{1}{2})^2 + \tfrac{1}{2} - 1 & \textit{Substitute } \tfrac{1}{2} \textit{ for } x. \\ &= 2(\tfrac{1}{4}) + \tfrac{1}{2} - 1 & \textit{Evaluate powers first.} \\ &= \tfrac{1}{2} + \tfrac{1}{2} - 1 & \textit{Multiply.} \\ &= 0 & \textit{Add and subtract terms.} \end{aligned}$$

Guidelines:

- A relation between x and y is a function of x if each value of x corresponds to exactly one value of y.

  To evaluate a function:
- Substitute the input value into the equation for the function.
- Simplify, using the order of operations listed on page 10 of the textbook.

EXERCISES

In 1–4, determine whether the relation is a function.

1. $\{(4,\ -2),\ (4,\ 2),\ (1,\ -1),\ (1,\ 1),\ (0,\ 0)\}$ **2.** $\{(-2,\ 3),\ (-1,\ 2),\ (0,\ 1),\ (1,\ -1),\ (2,\ -2)\}$

3. $\{(1,\ 3),\ (5,\ 7),\ (6,\ 8),\ (1,\ 4)\}$ **4.** $\{(-10,\ 46),\ (-5,\ 12),\ (0,\ 13),\ (5,\ 12),\ (10,\ 13)\}$

In 5–8, find the indicated value of the function.

5. $f(x) = x^2 + 9,\ \ f(-3)$ **6.** $f(x) = 4x^2 - 12x + 9,\ \ f(\tfrac{3}{2})$

7. $f(x) = -x^2 - 7x - 3,\ \ f(-1)$ **8.** $f(x) = -2x^2 + x - 1,\ \ f(-\tfrac{1}{2})$

Reteach
Chapter 6

Name _____

What you should learn :

6.2	How to perform operations with functions and how to find the domain of a function.

Correlation to Pupil's Textbook:

Mid-Chapter Test (p. 305)   **Chapter Test (p. 339)**
Exercises 7–14, 16–18   Exercises 2–4, 6

Examples   *Performing Functions Operations and Finding Domains*

a. Let $f(x) = -2x$ and $g(x) = x + 3$.

Sum

$h(x) = f(x) + g(x)$
$= -2x + (x + 3)$
$= -x + 3$

Difference

$h(x) = f(x) - g(x)$
$= -2x - (x + 3)$
$= -3x - 3$

Product

$h(x) = f(x) \cdot g(x)$
$= -2x(x + 3)$
$= -2x^2 - 6x$

Quotient

$h(x) = f(x) \div g(x)$
$= \dfrac{-2x}{x + 3}$

Composition $f \circ g$

$h(x) = f(g(x))$
$= f(x + 3)$
$= -2(x + 3)$
$= -2x - 6$

Composition $g \circ f$

$h(x) = g(f(x))$
$= g(-2x)$
$= -2x + 3$

b. $f(x) = -x^2 + 4x - 7$

$g(x) = \dfrac{5}{3x + 6}$
$3x + 6 \neq 0$
$x \neq -2$

$h(x) = \sqrt{x - 1}$
$x - 1 \geq 0$
$x \geq 1$

The domain of f is all real numbers.

The domain of g is all real numbers except −2. The number −2 is not in the domain because the denominator cannot be zero.

The domain of h is all real numbers that are greater than or equal to 1. The quantity under the radical must be nonnegative because the square root of a negative number is not a real number.

Guidelines:
- Functions can be added, subtracted, multiplied, and divided.
- The composition of the function f with the function g is $f(g(x))$.
- The domain of a function is the set of all x-values that make sense in the function's equation.

EXERCISES

In 1–8, let $f(x) = 2 - x$ and $g(x) = 3x$. Write and simplify an equation for $h(x)$. Then find the domain of h.

1. $h(x) = f(x) - g(x)$

2. $h(x) = f(x) \cdot g(x)$

3. $h(x) = f(x) \div g(x)$

4. $h(x) = 4g(x)$

5. $h(x) = f(x) + g(x)$

6. $h(x) = g(x) \div f(x)$

7. $h(x) = f(g(x))$

8. $h(x) = g(f(x))$

© D.C. Heath and Company

Algebra 2

Chapter 6 ▪ Functions **37**

Name _____

What you should learn :

| **6.3** | How to find inverse relations and functions, and how to verify that two functions are inverses. |

Correlation to Pupil's Textbook:

Mid-Chapter Test (p. 305)   **Chapter Test (p. 339)**
Exercises 15, 20   Exercises 5, 7–10

Examples   *Finding and Verifying Inverses*

a. To find the inverse of $\{(2, -1), (3, 7), (4, 15), (6, 31)\}$, switch the coordinates of each ordered pair in the relation to obtain

$$\{(-1, 2), (7, 3), (15, 4), (31, 6)\}.$$

b. Find the inverse of $f(x) = 2x + 1$.

$f(x) = 2x + 1$	*Original function*
$y = 2x + 1$	*Replace $f(x)$ by y.*
$x = 2y + 1$	*Switch x and y to obtain inverse.*
$y = \frac{1}{2}x - \frac{1}{2}$	*Solve the equation for y.*
$g(x) = \frac{1}{2}x - \frac{1}{2}$	*Replace y by $g(x)$.*

c. Verify that $f(x) = 2x + 1$ and $g(x) = \frac{1}{2}x - \frac{1}{2}$ are inverses of each other.

$f(g(x)) = f(\frac{1}{2}x - \frac{1}{2})$	*Substitute $\frac{1}{2}x - \frac{1}{2}$ for $g(x)$.*
$= 2(\frac{1}{2}x - \frac{1}{2}) + 1$	*Apply the formula for f.*
$= x - 1 + 1$	*Distributive Property*
$= x$	*Simplify.*
$g(f(x)) = g(2x + 1)$	*Substitute $2x + 1$ for $f(x)$.*
$= \frac{1}{2}(2x + 1) - \frac{1}{2}$	*Apply the formula for g.*
$= x + \frac{1}{2} - \frac{1}{2}$	*Distributive Property*
$= x$	*Simplify.*

Guidelines:
- To find the inverse of a function, replace $f(x)$ by y, switch x and y, and solve for y.
- To verify that f and g are inverses of each other, show that $f(g(x)) = x$ and that $g(f(x)) = x$.

EXERCISES

In 1–6, find the inverse of the function.

1. $f(x) = x - 3$ **2.** $f(x) = 3x - 2$ **3.** $f(x) = 5 - x$

4. $f(x) = 4x + 3$ **5.** $f(x) = 3 - 2x$ **6.** $f(x) = \frac{1}{2}x - 2$

In 7 and 8, verify that *f* and *g* are inverses of each other.

7. $f(x) = -4x + 1$, $g(x) = \frac{1}{4} - \frac{1}{4}x$ **8.** $f(x) = \frac{1}{2}x + 3$, $g(x) = 2x - 6$

Reteach

Chapter 6

Name _____

What you should learn:

| 6.4 | How to evaluate, graph, and write compound functions. |

Correlation to Pupil's Textbook:

Chapter Test (p. 339)
Exercises 11, 12

Examples *Evaluating, Graphing, and Writing Compound Functions*

a. Evaluate $f(3)$ and $f(0)$. Then sketch the graph of the compound function.

$$f(x) = \begin{cases} x - 1, & x < 1 \\ x^2 - 2x + 1, & x \geq 1 \end{cases}$$

Because $3 > 1$, use the second equation to evaluate $f(3)$.
Because $0 < 1$, use the first equation to evaluate $f(0)$.

$$f(3) = 3^2 - 2(3) + 1 = 4, \qquad f(0) = 0 - 1 = -1$$

To the left of $x = 1$, the graph is the line $y = x - 1$. To the right of $x = 1$, the graph is the parabola $y = x^2 - 2x + 1$, as shown at the right.

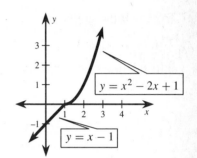

$y = x^2 - 2x + 1$

$y = x - 1$

b. Write the absolute value function $f(x) = |2x + 4|$ as a compound function.

$$|2x + 4| = \begin{cases} 2x + 4, & \text{if } 2x + 4 \geq 0 \\ -(2x + 4), & \text{if } 2x + 4 < 0 \end{cases} \qquad \textit{Definition of absolute value}$$

$$2x + 4 \geq 0 \quad \Rightarrow \quad x \geq -2 \qquad \textit{Solve first inequality for } x.$$

$$2x + 4 < 0 \quad \Rightarrow \quad x < -2 \qquad \textit{Solve second inequality for } x.$$

$$f(x) = \begin{cases} 2x + 4, & x \geq -2 \\ -2x - 4, & x < -2 \end{cases} \qquad \textit{Rewrite the compound function.}$$

Guidelines:
- Compound functions are defined by two or more equations.
- To graph a compound function, sketch the graph of each equation separately, then darken the portion that represent the graph of the function.
- To evaluate a compound function at a particular value of x, use the appropriate equation, as demonstrated in Example **a**.

EXERCISES

In 1–4, evaluate the function for the given value of x.

$$f(x) = \begin{cases} x - 5, & x < 2 \\ -x^2 + 1, & x \geq 2 \end{cases}$$

1. $f(-2)$ **2.** $f(0)$ **3.** $f(2)$ **4.** $f(5)$

5. Graph $f(x) = \begin{cases} x^2, & x < 0 \\ x, & x \geq 0 \end{cases}$ **6.** Graph $f(x) = \begin{cases} 1, & x < 0 \\ -x^2 + 2x + 1, & x \geq 0 \end{cases}$

In 7 and 8, write f(x) as a compound function.

7. $f(x) = |x - 3|$ **8.** $f(x) = |\frac{1}{2}x + 1|$

© D.C. Heath and Company

Algebra 2

Chapter 6 ▪ Functions **39**

Reteach
Chapter 6

Name _____

What you should learn :

6.5	How to sketch the graph of a function using translations and reflections.

Correlation to Pupil's Textbook:

Chapter Test (p. 339)
Exercise 13

Examples | *Using Translations and Reflections to Sketch a Graph*

a. To sketch the graph of

$$g(x) = |x| + 3,$$

shift the graph of $f(x) = |x|$ up 3 units.

b. To sketch the graph of

$$g(x) = -|x - 1|,$$

shift the graph of $f(x) = |x|$ to the right 1 unit. Then reflect the result in the x-axis.

c. To sketch the graph of

$$g(x) = (x + 3)^2 - 2,$$

shift the graph of $f(x) = x^2$ to the left 3 units and down 2 units.

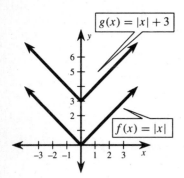

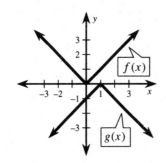

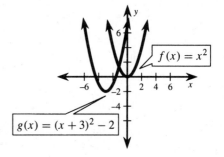

Guidelines: To sketch the graph of g using translations and reflections ($c > 0$):

- If $g(x) = f(x) + c$, shift graph of f *up* c units.
- If $g(x) = f(x) - c$, shift graph of f *down* c units.
- If $g(x) = f(x + c)$, shift graph of f *to the left* c units.
- If $g(x) = f(x - c)$, shift graph of f *to the right* c units.
- If $g(x) = -f(x)$, reflect graph of f in the x-axis.

EXERCISES

In 1–12, sketch the graph of the function.

1. $f(x) = |x| - 2$

2. $f(x) = |x| + 4$

3. $f(x) = |x - 3|$

4. $f(x) = -|x + 2|$

5. $f(x) = |x + 1| - 3$

6. $f(x) = |x - 2| - 2$

7. $f(x) = x^2 + 3$

8. $f(x) = x^2 - 5$

9. $f(x) = (x - 2)^2$

10. $f(x) = -(x - 2)^2$

11. $f(x) = (x - 1)^2 + 3$

12. $f(x) = (x + 2)^2 + 1$

Reteach

Chapter 6

Name _____

What you should learn :

| 6.6 | How to evaluate and classify recursive functions. |

Correlation to Pupil's Textbook:

Chapter Test (p. 339)
Exercises 14–16

| **Examples** | *Evaluating Recursive Functions and Finding Finite Differences* |

a. Find the first five values of the recursive function given by $f(1) = 1$ and
$f(n) = f(n-1) + 3$. Then, use the results to find the first differences.

$f(1) = 1$ *The value of $f(1)$ is given.*

$f(2) = f(1) + 3 = 1 + 3 = 4$ *Substitute 1 for $f(1)$ to find $f(2)$.*

$f(3) = f(2) + 3 = 4 + 3 = 7$ *Substitute 4 for $f(2)$ to find $f(3)$.*

$f(4) = f(3) + 3 = 7 + 3 = 10$ *Substitute 7 for $f(3)$ to find $f(4)$.*

$f(5) = f(4) + 3 = 10 + 3 = 13$ *Substitute 10 for $f(4)$ to find $f(5)$.*

The first differences are 3, 3, 3, and 3. These are found by subtracting
consecutive values of the function. $(4 - 1 = 3, 7 - 4 = 3,$ and so on.$)$

b. Because the first differences in Example **a** are all the same number, the
function has a linear model of the form $f(n) = an + b$.

$f(n) = an + b$ *General linear model*

$f(1) = a(1) + b = 1$ *Substitute 1 for n and 1 for $f(1)$.*

$f(2) = a(2) + b = 4$ *Substitute 2 for n and 4 for $f(2)$.*

$\begin{cases} a + b = 1 \\ 2a + b = 4 \end{cases}$ *Write system of linear equations in a and b.*

$a = 3$ and $b = -2$ *Solve system by substitution or linear combinations.*

$f(n) = 3n - 2$ *Linear model*

Guidelines: To classify a recursive function:
- If the first differences are all the same number, then the function has a linear model.
- If the second differences are all the same number, then the function has a quadratic model.

EXERCISES

**In 1–4, find the first seven values of the function. Then find a linear or
quadratic model for the function.**

1. $f(1) = 10$
$\quad f(n) = f(n-1) - 2$

2. $f(0) = 1$
$\quad f(n) = f(n-1) + 2n$

3. $f(1) = 2$
$\quad f(n) = f(n-1) + n - 1$

4. $f(1) = 3$
$\quad f(n) = f(n-1) + 2$

© D.C. Heath and Company

Chapter 6 ▪ *Functions* **41**

Reteach
Chapter 6

Name _____

What you should learn :

| 6.7 | How to find the mean, median, mode, and quartiles of a collection of numbers. |

Correlation to Pupil's Textbook:

Chapter Test (p. 339)
Exercise 18

| **Examples** | *Measuring Central Tendency* |

a. Find the mean, median, and mode of the following collection.

    15, 11, 19, 15, 14, 14, 13, 17, 11, 12, 17, 15, 14, 15

To begin, order the fourteen numbers.

    11, 11, 12, 13, 14, 14, 14, 15, 15, 15, 15, 17, 17, 19

To find the mean, divide the sum of the numbers by 14.

$$\text{Mean} = \frac{2(11) + 12 + 13 + 3(14) + 4(15) + 2(17) + 19}{14} \approx 14.4$$

The median is the average of the two middle numbers.

$$\text{Median} = \frac{14 + 15}{2} = 14.5$$

The mode is 15 because that is the number that occurs most frequently.

b. Find the quartiles of the collection in Example **a**. Then, sketch a box-and-whisker plot of the data.

$$\underbrace{11, \ 11, \ 12, \ 13, \ 14, \ 14, \ 14,}_{\text{Lower half}} \underbrace{15, \ 15, \ 15, \ 15, \ 17, \ 17, \ 19}_{\text{Upper half}}$$

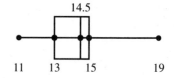

The first quartile is 13 (the median of the lower half).
The second quartile is 14.5 (the median of the entire collection).
The third quartile is 15 (the median of the upper half).
A box-and-whisker plot for the data labels the endpoints of the data and marks the quartiles. It is shown at the right.

Guidelines: To measure central tendency:
- Order the collection of data from smallest to largest.
- Find the mean, median, and mode as described on page 327 of the textbook.
- The quartiles can be found by dividing the ordered collection into two halves and then finding the medians.

EXERCISES

1. Find the mean, median, and mode of the following collection of scores on a test.

32, 72, 81, 95, 98, 58, 77, 75, 83, 97, 45, 89, 93, 57
82, 97, 52, 75, 79, 78, 99, 98, 54, 75, 85, 61, 55, 86

2. Find the first, second, and third quartiles of the collection of data in Exercise 1.

3. Construct a box-and-whisker plot of the collection of data in Exercise 1.

© D. C. Heath and Company

Name _____

What you should learn :

| | Correlation to Pupil's Textbook: |

7.1 How to use properties of exponents to evaluate and simplify exponential expressions.

Correlation to Pupil's Textbook:

Mid-Chapter Test (p. 367)	Chapter Test (p. 395)
Exercises 1–3, 7–9	Exercises 4, 21, 24

Examples | *Evaluating and Simplifying Exponential Expressions*

a. $2^3 \cdot 2^{-4} = 2^{3+(-4)}$

$\qquad = 2^{-1}$

$\qquad = \frac{1}{2}$

b. $(2 \cdot 3^{-2})^{-3} = 2^{-3} \cdot (3^{-2})^{-3}$

$\qquad = 2^{-3} \cdot 3^6$

$\qquad = \frac{1}{2^3} \cdot 3^6$

$\qquad = \frac{729}{8}$

c. $\dfrac{2x^{-3}y}{xy^{-2}} = \dfrac{2y \cdot y^2}{x \cdot x^3}$

$\qquad = \dfrac{2y^3}{x^4}$

d. Solve for x in the equation $\dfrac{x^{11}y^2}{3x^8y} \cdot \dfrac{4}{x^2y} = \dfrac{2}{9}$.

$\dfrac{x^{11}y^2}{3x^8y} \cdot \dfrac{4}{x^2y} = \dfrac{2}{9}$ *Rewrite original equation.*

$\dfrac{(x^{11}y^2)(4)}{(3x^8y)(x^2y)} = \dfrac{2}{9}$ *Multiply fractions.*

$\dfrac{4x^{11}y^2}{3x^{10}y^2} = \dfrac{2}{9}$ *Product of Powers Property.*

$\dfrac{4x}{3} = \dfrac{2}{9}$ *Quotient of Powers Property*

$x = \dfrac{1}{6}$ *Multiply both sides by $\frac{3}{4}$.*

Guidelines: To simplify exponential expressions:
- Use the properties of exponents listed on page 346 of the textbook.
- Write the answers without negative exponents.

EXERCISES

In 1–9, simplify the expression.

1. $(2)^6 \cdot (2)^{-3}$

2. $(-2 \cdot 3^2)^3$

3. $t^3 \cdot t^2$

4. $(2^2)^{-2}$

5. $(3x^2y)^{-2}$

6. $(6)^{-2} \cdot (3)^0 \cdot (2)^3$

7. $\dfrac{x^{-3}y^2}{xy^{-3}}$

8. $\dfrac{-4y^8}{y^2} \cdot \dfrac{3x^8}{8y}$

9. $\dfrac{-2x^{-3}}{y^3} \cdot \dfrac{x^3}{xy^2}$

In 10–12, solve for x.

10. $\dfrac{3^2}{3^x} = 3^5$

11. $\dfrac{x^3}{2} \cdot \dfrac{4}{x^2} = 10$

12. $\dfrac{x^3}{x^5} = 16$

Scrambled answers for first column of exercises: $\dfrac{y^5}{x^4}, -3, 8, \frac{1}{16}$

Name _____

What you should learn :

| **7.2** | How to use exponential formulas and sketch graphs of exponential equations. |

Correlation to Pupil's Textbook:

Mid-Chapter Test (p. 367)   Chapter Test (p. 395)
Exercises 10, 11, 15–20   Exercises 22, 23

Examples   *Using Exponential Formulas and Graphing Exponential Equations*

a. How much money must be deposited in an account that pays 7% annual interest, compounded quarterly, to have a balance of $4000 after five years?

$$A = P\left(1 + \frac{r}{n}\right)^{nt}$$ *Formula for compound interest*

$$4000 = P\left(1 + \frac{0.07}{4}\right)^{(4)(5)}$$ *Let $A = 4000$, $r = 0.07$, $n = 4$, and $t = 5$.*

$$4000 = P(1.0175)^{20}$$ *Simplify.*

$$4000 \approx P(1.4148)$$ *Use a calculator.*

$$2827.30 \approx P$$ *Divide both sides by 1.4148.*

You should deposit $2827.30.

b. Sketch the graph of $y = 2(\frac{1}{3})^x$.

Begin by making a table of values.

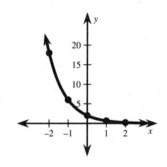

x	-2	-1	0	1	2
y	18	6	2	$\frac{2}{3}$	$\frac{2}{9}$

Plot the points indicated by the table. Then connect the points with a smooth curve, as shown at the right.

Guidelines:
- The formula for compound interest is $A = P(1 + \frac{r}{n})^{nt}$.

    To sketch the graph of an exponential function:

- Make a table of values.
- Plot the points.
- Connect the points with a smooth curve.

EXERCISES

1. $500 is invested at 6% annual interest, compounded quarterly. What is the balance after three years?

2. $2000 is invested at 7% annual interest, compounded monthly. What is the balance after five years?

3. How much money must be deposited in an account that pays 9% annual interest, compounded monthly, to have a balance of $1000 after three years?

4. Sketch the graph of $y = 2^x$.

5. Sketch the graph of $y = (\frac{1}{3})^x$.

© D. C. Heath and Company

Name _____

What you should learn :

| 7.3 | How to evaluate the nth root of a real number. |

Correlation to Pupil's Textbook:

Mid-Chapter Test (p. 367)
Exercises 4–6, 12–14

Examples   *Evaluating the nth Roots of Real Numbers*

a. To evaluate $16^{3/4}$, use the fact that 2 is a 4th root of 16.

$$16^{3/4} = (16^{1/4})^3$$
$$= 2^3$$
$$= 8$$

b. To evaluate $27^{-4/3}$, use the fact that 3 is a cube root of 27.

$$27^{-4/3} = \frac{1}{27^{4/3}}$$
$$= \frac{1}{(27^{1/3})^4}$$
$$= \frac{1}{3^4}$$
$$= \tfrac{1}{81}$$

c. To evaluate $(-32)^{2/5}$, use the fact that -2 is a 5th root of -32.

$$(-32)^{2/5} = ((-32)^{1/5})^2$$
$$= (-2)^2$$
$$= 4$$

d. The volume of a basketball is 1767.15 cubic inches. Find the radius of the basketball.

$$V = \tfrac{4}{3}\pi r^3 \qquad \textit{Formula for volume of a sphere}$$

$$\frac{3V}{4\pi} = r^3 \qquad \textit{Divide both sides by } \tfrac{4}{3}\pi.$$

$$\left(\frac{3V}{4\pi}\right)^{1/3} = r \qquad \textit{Take cube root of both sides.}$$

$$\left(\frac{3(1767.15)}{4\pi}\right)^{1/3} = r \qquad \textit{Substitute 1767.15 for V.}$$

$$7.5 \approx r \qquad \textit{Use a calculator.}$$

The radius of the basketball is about 7.5 inches.

Guidelines: To evaluate the expression $a^{m/n}$:
- Rewrite the expression as $(a^{1/n})^m$.
- Find the *n*th root of *a*.
- Raise that quantity to the *m*th power.

EXERCISES

In 1–6, evaluate the expression without using a calculator.

1. $27^{2/3}$

2. $343^{1/3}$

3. $(-32)^{3/5}$

4. $81^{-3/4}$

5. $16^{3/2}$

6. $36^{-3/2}$

In 7–9, rewrite the expression using rational exponents.

7. $\sqrt{7}$

8. $\sqrt[3]{-17}$

9. $\sqrt[9]{180}$

Name _____

What you should learn :

| 7.4 | How to evaluate and simplify radical expressions and rational-exponent expressions. |

Correlation to Pupil's Textbook:

Chapter Test (p. 395)
Exercises 7–12

Examples *Using Properties of Roots*

a. $2 \cdot 2^{1/2} = 2^1 \cdot 2^{1/2}$
$\qquad = 2^{(1+1/2)}$
$\qquad = 2^{3/2}$

b. $(5^{1/4})^{16/3} = 5^{(1/4)(16/3)}$
$\qquad = 5^{4/3}$

c. $(-\frac{3}{4})^{7/3}(-\frac{3}{4})^{2/3} = (-\frac{3}{4})^{(7/3+2/3)}$
$\qquad = (-\frac{3}{4})^3$
$\qquad = -\frac{27}{64}$

d. $\sqrt[3]{50} \cdot \sqrt[3]{20} = \sqrt[3]{50 \cdot 20}$ *Property 3 on page 368 of the textbook.*
$\qquad\quad = \sqrt[3]{1000}$ *Simplify.*
$\qquad\quad = 10$ *Find cube root of* 1000.

e. $\sqrt{27} - \sqrt{3} = \sqrt{9 \cdot 3} - \sqrt{3}$ *Substitute* $9 \cdot 3$ *for* 27.
$\qquad\quad\;\; = \sqrt{9} \cdot \sqrt{3} - \sqrt{3}$ *Property 3 on page 368 of the textbook.*
$\qquad\quad\;\; = 3\sqrt{3} - \sqrt{3}$ *Substitute* 3 *for* $\sqrt{9}$.
$\qquad\quad\;\; = 2\sqrt{3}$ *Simplify.*

f. $\dfrac{\sqrt[3]{375}}{\sqrt[3]{3}} = \sqrt[3]{\dfrac{375}{3}}$ *Property 5 on page 368 of the textbook.*
$\qquad\quad = \sqrt[3]{125}$ *Simplify.*
$\qquad\quad = 5$ *Find cube root of* 125.

Guidelines: To evaluate and simplify expression containing radicals and rational exponents:
- Use the properties of roots listed on page 368 of the textbook.

EXERCISES

In 1–15, simplify the expression.

1. $(2) \cdot (2)^{1/2}$ **2.** $\dfrac{6^{4/5}}{6^{1/5}}$ **3.** $(3^{2/3})^{3/4}$ **4.** $(8x^6)^{1/3}$

5. $\dfrac{72^{1/2}}{2^{1/2}}$ **6.** $64^{-1/3}$ **7.** $\sqrt[3]{16} + \sqrt[3]{2}$ **8.** $\sqrt[4]{4} \cdot \sqrt[4]{4}$

9. $\dfrac{\sqrt{98}}{\sqrt{2}}$ **10.** $\dfrac{3^{1/2}}{3^{1/3}}$ **11.** $\sqrt[5]{-32y^5}$ **12.** $\sqrt{36x^{16}}$

13. $\sqrt[4]{256x^8 y}$ **14.** $\sqrt{2x}\sqrt{8x}$ **15.** $(5^{1/2} \cdot 5^{1/4})^3$

© D.C. Heath and Company

Name _____

What you should learn :

| 7.5 | How to solve equations that have radicals and rational exponents. |

Correlation to Pupil's Textbook:

Chapter Test (p. 395)
Exercises 1–3, 13–20

Examples *Solving Radical Equations*

a.

$(x^2 - 2x + 5)^{2/3} = 4$	*Original equation*
$((x^2 - 2x + 5)^{2/3})^{3/2} = 4^{3/2}$	*Raise both sides to the $\frac{3}{2}$ power.*
$x^2 - 2x + 5 = 8$	*Apply properties of exponents.*
$x^2 - 2x - 3 = 0$	*Write the quadratic equation in standard form.*
$x = 3, -1$	*Apply the quadratic formula.*

The solutions are 3 and -1. Check these in the original equation.

b.

$\sqrt[3]{2x + 4} - 2\sqrt[3]{3 - x} = 0$	*Original equation*
$\sqrt[3]{2x + 4} = 2\sqrt[3]{3 - x}$	*Isolate one radical on the left side of the equation.*
$(\sqrt[3]{2x + 4})^3 = (2\sqrt[3]{3 - x})^3$	*Cube both sides of the equation.*
$(\sqrt[3]{2x + 4})^3 = 2^3(\sqrt[3]{3 - x})^3$	*Apply properties of exponents.*
$2x + 4 = 8(3 - x)$	*Apply properties of exponents.*
$2x + 4 = 24 - 8x$	*Distributive Property*
$10x = 20$	*Isolate the terms containing x.*
$x = 2$	*Divide both sides by 10.*

The solution is 2. Check this in the original equation.

Guidelines: To solve a radical equation:
- Isolate the term involving the radical on one side of the equation.
- Raise both sides of the equation to the appropriate power to eliminate the radical.
- Solve the resulting equation.
- Check each solution in the original equation.

EXERCISES

In 1–12, solve the equation. Check for extraneous solutions.

1. $x^{3/4} = 8$ **2.** $x^{5/4} = 32$ **3.** $(x^2 + 4)^{2/3} = 25$

4. $\sqrt{4 + 3x} = 10$ **5.** $\sqrt{2x + 1} = 7$ **6.** $\sqrt{x} + x = 2$

7. $\sqrt{x} + 2 = x$ **8.** $\sqrt{x + 4} + 2 = x$ **9.** $3x + 5 = \sqrt{2 - 2x}$

10. $\sqrt[3]{4x - 1} = 3$ **11.** $\sqrt{2 - 5x} = 5x$ **12.** $\sqrt[4]{x + 1} + \sqrt[4]{5x - 7} = 0$

Scrambled answers for first column of exercises: 16, 4, 32, 7

Reteach

Chapter 7

What you should learn :

7.6	How to graph square root functions and cube root functions.

Correlation to Pupil's Textbook:

Chapter Test (p. 395)
Exercises 5, 6

Examples *Graphing Radical Functions*

a. To sketch the graph of

$$g(x) = \sqrt{x+1} - 2,$$

shift the graph of $f(x) = \sqrt{x}$ one unit to the left and two units down.

Domain of g: $x \geq -1$
Range of g: $y \geq -2$

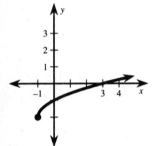

b. To sketch the graph of

$$g(x) = \sqrt[3]{x-3},$$

shift the graph of $f(x) = \sqrt[3]{x}$ three units to the right.

Domain of g: All real numbers
Range of g: All real numbers

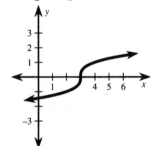

c. To sketch the graph of

$$g(x) = -\sqrt{x+4},$$

shift the graph of $f(x) = \sqrt{x}$ four units to the left. Then reflect the result in the x-axis.

Domain of g: $x \geq -4$
Range of g: $y \leq 0$

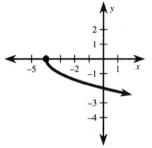

Guidelines: To sketch functions of the form $y = \sqrt{x-a} + b$ or $y = \sqrt[3]{x-a} + b$:
- Know the graphs of $f(x) = \sqrt{x}$ and $f(x) = \sqrt[3]{x}$.
- Use translations and reflections as discussed in Lesson 6.5 in the text.

EXERCISES

In 1–9, sketch the graph of the function. State the domain and range of the function.

1. $f(x) = \sqrt{x+3}$

2. $f(x) = \sqrt{x} + 3$

3. $f(x) = -\sqrt{x+3}$

4. $f(x) = \sqrt{x+3} + 3$

5. $f(x) = \sqrt[3]{x} - 1$

6. $f(x) = \sqrt[3]{x+2}$

7. $f(x) = \sqrt[3]{x-1} + 2$

8. $f(x) = -\sqrt[3]{x-3}$

9. $f(x) = -\sqrt{x} + 2$

Reteach
Chapter 8

Name _____

What you should learn :

| 8.1 | How to graph exponential functions and write exponential growth and decay models. |

Correlation to Pupil's Textbook:

Mid-Chapter Test (p. 419)   **Chapter Test (p. 453)**
Exercises 1–3, 30, 31   Exercise 1

Examples | *Graphing and Using Exponential Functions*

a. To sketch the graph of $g(x) = 3^{x-1}$, shift the graph of $f(x) = 3^x$ one unit to the right. The graph of g is shown at the right.

Graph of $g(x) = 3^{x-1}$

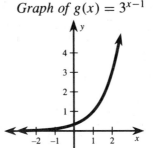

Graph of $h(x) = -(\frac{1}{3})^{x+2} - 1$

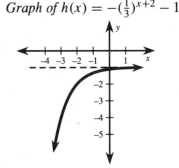

b. To sketch the graph of $h(x) = -(\frac{1}{3})^{x+2} - 1$, shift the graph of $f(x) = (\frac{1}{3})^x$ two units to the left. Reflect the result in the x-axis. Then shift the graph one unit down. The graph of h has a horizontal asymptote at $y = -1$, as shown at the right.

c. A diamond ring was purchased twenty years ago for \$500. The value of the ring increased by 8% each year. What is the value of the ring today?

Let V be the present value. In the exponential growth model $V = Ca^t$, the initial value is $C = 500$, the percent of previous value is $a = 1.08$, and the time is $t = 20$ years.

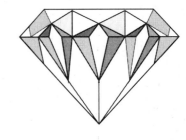

$V = Ca^t$ *Exponential growth model*

$V = 500(1.08)^{20}$ *Substitute $C = 500$, $a = 1.08$, and $t = 20$.*

$V \approx 2330.48$ *Use a calculator.*

The value of the ring is about \$2330.48.

Guidelines: To sketch the exponential function $y = -a^{x+b} + c$:
- Shift the graph of $y = a^x$ horizontally $|b|$ units.
- Reflect the result in the x-axis.
- Shift the result vertically $|c|$ units.
- The graph has a horizontal asymptote at $y = c$.

EXERCISES

In 1–6, sketch the graph of the function.

1. $f(x) = 3^x + 1$ **2.** $f(x) = (\frac{1}{2})^{x-1}$ **3.** $f(x) = -3^{x-1}$

4. $f(x) = -2^x + 2$ **5.** $f(x) = 2^{x+2} - 1$ **6.** $f(x) = -(\frac{1}{2})^{x-2} + 1$

7. A compact disk player is purchased for \$429. Each year the value of the player decreases by 5%. Find the value of the compact disk player five years from now.

Reteach
Chapter 8

Name _____

Examples   *Evaluating and Graphing Logarithmic Functions*

a. Evaluate $\log_{27} 3$ without using a calculator.

$x = \log_{27} 3$	*Set x equal to the given expression.*
$27^x = 3$	*Rewrite in exponential form.*
$(3^3)^x = 3$	*To obtain equal bases, substitute 3^3 for 27.*
$3^{3x} = 3^1$	*Apply properties of exponents.*
$3x = 1$	*Because the bases are equal, the exponents must be equal.*
$x = \frac{1}{3}$	*Divide both sides by 3.*

b. You can solve the equation $\log_x 81 = 4$ as follows.

$$\log_x 81 = 4 \;\Rightarrow\; x^4 = 81 \;\Rightarrow\; x^4 = 3^4 \;\Rightarrow\; x = 3$$

c. Sketch the graph of $f(x) = \log_3 x + 1$.

Begin by constructing a table of values, choosing only positive values of x.

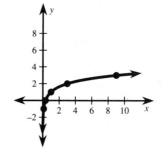

x	$\frac{1}{9}$	$\frac{1}{3}$	1	3	9
f(x)	-1	0	1	2	3

Plot the points from the table and connect them with a smooth curve, as shown at the right.

Guidelines: To evaluate a logarithmic expression without using a calculator:
- Set the expression equal to x.
- Rewrite the equation in exponential form.
- Solve the equation for x.

EXERCISES

In 1–3, evaluate the expression without using a calculator.

1. $\log_7 7$ **2.** $\log_5 25$ **3.** $\log_4 1$ **4.** $\log_2 \frac{1}{16}$ **5.** $\log_{16} 4$ **6.** $\log_3 81$

In 7–9, solve the equation for **x**.

7. $\log_2 64 = x$ **8.** $\log_x 9 = 2$ **9.** $\log_{16} x = \frac{1}{2}$

In 10–12, sketch the graph of the function.

10. $f(x) = \log_2 x - 2$ **11.** $f(x) = \log_2(x + 1)$ **12.** $f(x) = \log_3 x$

Name _____

What you should learn :

8.3	How to use properties of logarithms to condense and expand logarithmic expressions.

Examples *Using Properties of Logarithms*

a. Expand the expression $\log_{10} \sqrt{\frac{10}{t}}$.

$$\log_{10} \sqrt{\tfrac{10}{t}} = \log_{10} \left(\tfrac{10}{t}\right)^{1/2} \qquad \textit{Rewrite the square root as a power.}$$
$$= \tfrac{1}{2} \log_{10} \left(\tfrac{10}{t}\right) \qquad \textit{Power Property}$$
$$= \tfrac{1}{2} \left(\log_{10} 10 - \log_{10} t\right) \qquad \textit{Quotient Property}$$
$$= \tfrac{1}{2} \left(1 - \log_{10} t\right) \qquad \textit{Special logarithmic value: } \log_{10} 10 = 1$$

b. Condense the expression $2 \log_6 x + \tfrac{1}{2} \log_6 z$.

$$2 \log_6 x + \tfrac{1}{2} \log_6 z = \log_6 x^2 + \log_6 z^{1/2} \qquad \textit{Power Property}$$
$$= \log_6 (x^2 z^{1/2}) \qquad \textit{Product Property}$$
$$= \log_6 x^2 \sqrt{z} \qquad \textit{Rewrite } \tfrac{1}{2} \textit{ power as a square root.}$$

c. Solve $\log_2 \tfrac{8}{3} = x - \log_2 3$ for x.

$$\log_2 \tfrac{8}{3} = x - \log_2 3 \qquad \textit{Original equation}$$
$$\log_2 8 - \log_2 3 = x - \log_2 3 \qquad \textit{Quotient Property}$$
$$\log_2 8 = x \qquad \textit{Simplify.}$$
$$2^x = 8 \qquad \textit{Rewrite in exponential form.}$$
$$x = 3 \qquad \textit{Solve for } x.$$

Guidelines: To condense or expand a logarithmic expression, use the following:
- $\log_a(uv) = \log_a u + \log_a v$
- $\log_a \dfrac{u}{v} = \log_a u - \log_a v$
- $\log_a u^n = n \log_a u$

EXERCISES

In 1–6, expand the expression.

1. $\log_2 3x^2$

2. $\log_4 4y$

3. $\log_3 \tfrac{1}{3}$

4. $\log_{10} xyz$

5. $\log_5 \dfrac{x^2}{y}$

6. $\log_2 \dfrac{\sqrt{x}}{yz}$

In 7–10, condense the expression.

7. $\log_5 3 + 2 \log_5 x$

8. $\log_{10} y - 3 \log_{10} 2$

9. $5 \log_4 x - 2 \log_4 y + 3 \log_4 z$

Name _____

What you should learn :

8.4	How to use the number e as a base of an exponential function.

Correlation to Pupil's Textbook:

Chapter Test (p. 453)
Exercises 4–6, 21, 22

Examples *Using the Natural Base e*

a. $\dfrac{8e^8}{2e^2} = 4e^{8-2} = 4e^6$

b. $(2e^{-0.6})^{-5} = 2^{-5}(e^{-0.6})^{-5} = \dfrac{1}{2^5}(e^3) = \dfrac{e^3}{32}$

c. Sketch the graph of $f(x) = e^x - 2$.

Begin by constructing a table of values. Use a calculator to evaluate $f(x)$.

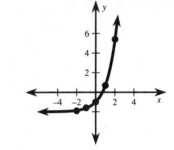

x	−2	−1	0	1	2
f(x)	−1.9	−1.6	−1	0.7	5.4

Plot the points from the table and connect them with a smooth curve, as shown at the right.

d. How much money must be deposited in an account that pays 7% annual interest, compounded continuously, to have a balance of $4000 after five years?

$A = Pe^{rt}$	*Formula for continuous compounding*
$4000 = Pe^{(0.07)(5)}$	*Substitute 4000 for A, 0.07 for r, and 5 for t.*
$\dfrac{4000}{e^{0.35}} = P$	*Divide both sides by $e^{0.35}$.*
$\$2818.75 \approx P$	*Use a calculator. You should deposit $2818.75.*

Guidelines: To use the natural base e:
- All the properties for exponents apply to the natural base e.
- The balance for continuous compounding is $A = Pe^{rt}$.

EXERCISES

In 1–6, simplify the expression.

1. $e^3 \cdot (e^4)^2$

2. $\dfrac{e^6}{e^3}$

3. $2(e^2)^{-1}$

4. $(3e^2)^3$

5. $\sqrt{16e^{36}}$

6. $e^3 \cdot e^{x-1} \cdot e^x$

In 7 and 8, sketch the graph of the function.

7. $f(x) = e^{2x} + 1$

8. $f(x) = e^{-2x} + 1$

9. $500 is invested at 6% annual interest, compounded continuously. What is the balance after three years?

10. How much must be deposited in an account that pays 9% annual interest, compounded continuously, to have a balance of $1000 after five years?

Reteach
Chapter 8

Name _____

Correlation to Pupil's Textbook:

Chapter Test (p. 453)
Exercises 2, 7, 10, 12, 14, 15

Examples | *Evaluating and Graphing Natural Logarithmic Functions*

a. $\ln 13\sqrt{x} = \ln 13 + \ln \sqrt{x} = \ln 13 + \ln x^{1/2} = \ln 13 + \frac{1}{2}\ln x$

b. $2\ln x + 3\ln y - \frac{1}{2}\ln z = \ln x^2 + \ln y^3 - \ln \sqrt{z} = \ln \dfrac{x^2 y^3}{\sqrt{z}}$

c. $\ln \dfrac{1}{e^2} = \ln 1 - \ln e^2 = \ln 1 - 2\ln e = 0 - 2(1) = -2$

d. $\ln 5e^3 = \ln 5 + \ln e^3 = \ln 5 + 3\ln e = \ln 5 + 3(1) = \ln 5 + 3$

e. Sketch the graph of
$$g(x) = \ln(x + 1) + 3.$$

To sketch the graph of g, you can shift the graph of $f(x) = \ln x$ one unit to the left. Then shift the result up 3 units. The line $x = -1$ is a vertical asymptote of the graph, as shown at the right.

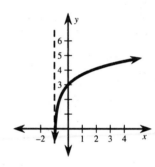

Guidelines: • To evaluate a natural logarithmic expression, use the properties of logarithms.

To graph a natural logarithmic function, use one of the following:
• Use transformations as shown in Example **e** above.
• Make a table of values, plot the points and connect them with a smooth curve.

EXERCISES

In 1–6, expand the expression.

1. $\ln \sqrt[4]{x}$ **2.** $\ln 2x^3 y^6$ **3.** $\ln \sqrt[3]{\dfrac{a^2 b}{c}}$ **4.** $\ln \dfrac{3y^2}{x}$ **5.** $\ln \dfrac{x}{yz}$ **6.** $\ln(2x^3)^2$

In 7–9, condense the expression.

7. $\ln 3 + 4\ln x$ **8.** $\ln x + \frac{1}{2}\ln y - \ln z$ **9.** $2(\ln x + \ln y) - \ln z$

In 10–12, simplify the expression.

10. $\ln e^{17}$ **11.** $\ln \dfrac{1}{e}$ **12.** $\ln 6e^2$

In 13–15, sketch the graph of the function.

13. $f(x) = \ln(x - 1)$ **14.** $f(x) = \ln x + 2$ **15.** $f(x) = \ln 1 - \ln x$

Reteach
Chapter 8

What you should learn :

8.6 How to solve exponential and logarithmic equations.	**Correlation to Pupil's Textbook:** Chapter Test (p. 453) Exercises 16, 17, 25

Examples	*Solving Exponential and Logarithmic Equations*

a.

$10^{2x} - 6 = 146$	*Original equation*
$10^{2x} = 152$	*To isolate the term containing x, add 6 to both sides.*
$\log_{10} 10^{2x} = \log_{10} 152$	*Take log with base 10 of both sides.*
$2x = \log_{10} 152$	*Simplify, using special logarithmic property, $\log_a a^x = x$.*
$x = \frac{1}{2} \log_{10} 152 \approx 1.091$	*Divide both sides by 2.*

b.

$\frac{1}{2} e^{-2x} = 6$	*Original equation*
$e^{-2x} = 12$	*Multiply both sides by 2.*
$\ln e^{-2x} = \ln 12$	*Take natural log of both sides.*
$-2x = \ln 12$	*Simplify.*
$x = -\frac{1}{2} \ln 12 \approx -1.242$	*Divide both sides by -2.*

c.

$3 \ln 5x - 2 = 16$	*Original equation*
$3 \ln 5x = 18$	*To isolate the term containing x, add 2 to both sides.*
$\ln 5x = 6$	*Divide both sides by 3.*
$e^{\ln 5x} = e^6$	*Exponentiate both sides.*
$5x = e^6$	*Simplify.*
$x = \frac{1}{5} e^6 \approx 80.686$	*Divide both sides by 5.*

Guidelines: To solve exponential or logarithmic equations:
- Isolate the term containing the variable on one side of the equation.
- Divide both sides by the coefficient of the variable term.
- If solving an exponential equation, take the log of both sides. If solving a logarithmic equation, exponentiate both sides.
- Simplify, using $\log_a a^x = x$ or $a^{\log_a x} = x$.

EXERCISES

In 1–12, solve for x.

1. $3^x = 8.4$ **2.** $e^{2x} = 17$ **3.** $2^x + 1 = 5$ **4.** $e^{-x} - 6 = 9$

5. $\frac{1}{2} e^{3x} + 11 = 20$ **6.** $0.5(10)^x - 4.6 = 3.7$ **7.** $9 - 4e^x = 5$ **8.** $2 - 3 \ln x = 1$

9. $4 \ln 2x = 5$ **10.** $4 \log_3 3x = 20$ **11.** $7 + 13 \log_{10} x = 5$ **12.** $7 - 2 \ln x = 1$

Scrambled answers for first column of exercises: 0.963, 1.937, 1.745

Reteach
Chapter 8

Name _____

What you should learn :

| **8.7** | How to graph a logistics growth function. |

Correlation to Pupil's Textbook:

Chapter Test (p. 453)
Exercises 3, 23, 24

Examples | Graphing and Using Logistics Growth Functions

a. Sketch the graph of $f(x) = \dfrac{3}{1 + e^{-0.7x}} + 2$.

Begin by constructing a table of values. Use a calculator to evaluate $f(x)$.

x	-3	-2	-1	0	1	2	3
$f(x)$	2.3	2.6	3.0	3.5	4.0	4.4	4.7

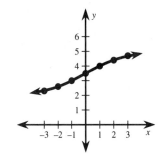

Plot the points from the table and connect them with a smooth curve, as shown at the right. Notice that $y = 2$ and $y = 5$ are horizontal asymptotes. The range of the function is $2 < y < 5$.

b. In a community of 50,000 people, the number of people, y, exposed to an advertisement is modeled by the logistics equation

$$y = \frac{50{,}000}{1 + 99e^{-1.4t}}$$

where t is the time in days. How many people in the community will have seen the advertisement after 3 days?

$$y = \frac{50{,}000}{1 + 99e^{-1.4(3)}} = \frac{50{,}000}{1 + 99e^{-4.2}} \approx 20{,}124$$

About 20,124 people will have seen the advertisement.

Guidelines:

To sketch the graph of the logistics growth function, $y = \dfrac{a}{1 + be^{-ct}} + d$:

- Construct a table of values.
- Plot the points and connect them with a smooth curve.
- $y = d$ and $y = a + d$ are horizontal asymptotes of the graph.

EXERCISES

In 1–6, sketch the graph of the function.

1. $f(x) = \dfrac{2}{1 + 3e^{-x}}$

2. $f(x) = \dfrac{3}{1 + e^{-0.2x}}$

3. $f(x) = \dfrac{50}{1 + 49e^{-2x}}$

4. $f(x) = \dfrac{1}{1 + e^{-x}} + 1$

5. $f(x) = \dfrac{2}{1 + 2e^{-3x}} + 3$

6. $f(x) = \dfrac{3}{1 + e^{-1.1x}} - 2$

In 7 and 8, evaluate the function at the given value of t.

7. $y = \dfrac{300}{1 + e^{-2t}}, \quad t = 2$

8. $y = \dfrac{5}{1 + e^{-0.3t}}, \quad t = 10$

Name _____

What you should learn :

| **9.1** How to add, subtract, and multiply polynomials. |

Correlation to Pupil's Textbook:

Mid-Chapter Test (p. 487)   **Chapter Test (p. 513)**
Exercises 1–4, 19, 21   Exercises 1, 2

| **Examples** | *Adding, Subtracting, and Multiplying Polynomials*

a. $(-x^2 + 4x + 6) - (x^2 - 3x + 1) = -x^2 + 4x + 6 - x^2 + 3x - 1$ *Distribute.*

$$= -2x^2 + 7x + 5 \qquad \text{\textit{Collect like terms.}}$$

b. $(x - 2)(2x^2 - 2x + 1) = (x - 2)(2x^2) - (x - 2)(2x) + (x - 2)(1)$ *Distribute.*

$$= 2x^3 - 4x^2 - 2x^2 + 4x + x - 2 \qquad \text{\textit{Distribute.}}$$

$$= 2x^3 - 6x^2 + 5x - 2 \qquad \text{\textit{Collect like terms.}}$$

c. $(2x + 7)^2 = (2x)^2 + 2(2x)(7) + (7)^2$ *Square of a binomial*

$$= 4x^2 + 28x + 49 \qquad \text{\textit{Simplify.}}$$

d. $(x - 5)^3 = (x)^3 - 3(x)^2(5) + 3(x)(5)^2 - (5)^3$ *Cube of a binomial*

$$= x^3 - 15x^2 + 75x - 125 \qquad \text{\textit{Simplify.}}$$

e. To find the area of the shaded region at the right, subtract the area of the smaller rectangle from the area of the larger rectangle.

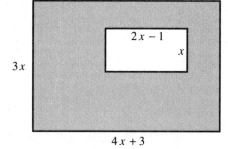

$3x$

$4x + 3$

$$\text{Area} = 3x(4x + 3) - x(2x - 1)$$

$$= 12x^2 + 9x - 2x^2 + x$$

$$= 10x^2 + 10x$$

Guidelines: To multiply polynomials:
- Use the Distributive Property.
- For two binomials, use the FOIL pattern.
- Know the formulas for cubing a binomial.

EXERCISES

In 1–10, perform the indicated operation.

1. $(7x^3 + x^2 - 2x + 1) - (16x^2 - 7)$ **2.** $(5x^2 - 6x - 1) - (4x^2 - 2x + 1)$

3. $3x(x^2 + x - 2)$ **4.** $-2x(1 - x - x^2)$

5. $(2x - 5)(x + 2)$ **6.** $(3 - 2x)^2$

7. $(x + 3)(x^2 - 2x - 1)$ **8.** $(x - 2)(x^2 + 2x + 4)$

9. $(x - 5)(x + 5)$ **10.** $(2x - 1)^3$

Name _____

What you should learn :

| 9.2 | How to sketch the graph of a polynomial function. |

Correlation to Pupil's Textbook:

Mid-Chapter Test (p. 487)   **Chapter Test (p. 513)**
Exercises 5, 6   Exercises 3, 4

Examples   *Sketching the Graphs of Polynomial Functions*

a. Sketch the graph of

$f(x) = x^3 - 3x^2 + 2x - 1$.

Because the leading coefficient is positive, the graph rises to the right. Because the degree of the polynomial is odd (and the graph rises to the right), the graph falls to the left. Because the degree of the polynomial is 3, it has at most two turns. The y-intercept is $(0, -1)$.

b. Sketch the graph of

$f(x) = -x^4 - 2x^3 + 3x^2 + 3x + 4$.

Because the leading coefficient is negative, the graph falls to the right. Because the degree of the polynomial is even (and the graph falls to the right), the graph falls to the left. Because the degree of the polynomial is 4, it has at most three turns. The y-intercept is $(0, 4)$.

c. Sketch the graph of

$g(x) = -(x - 1)^4 - 3$.

To sketch the graph of g, begin by sketching the graph of $f(x) = x^4$. Then shift the graph of f one unit to the right and reflect the result in the x-axis. Finally, shift the graph down three units. The y-intercept is $(0, -4)$.

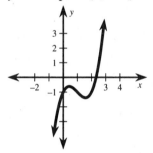

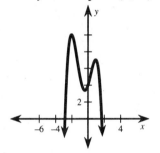

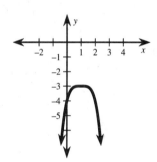

Guidelines: To sketch the graph of a polynomial function:
- Know the graphs of $f(x) = x^2$, $f(x) = x^3$, $f(x) = x^4$, and $f(x) = x^5$. Then use translations and reflections as demonstrated in part **c** above.
- A function of degree n has at most $n - 1$ turns.
- If the leading coefficient is positive, the graph rises to the right. If it is negative, the graph falls to the right.
- If the degree is even, then the graph has the same behavior left and right.

EXERCISES

In 1–12, sketch the graph of the function.

1. $f(x) = x^4 - 1$

2. $f(x) = -x^5 + 1$

3. $f(x) = -x^4 - 1$

4. $f(x) = x^5 + 2$

5. $f(x) = (x - 1)^4$

6. $f(x) = (x - 2)^5$

7. $f(x) = -(x + 2)^3$

8. $f(x) = (x + 1)^5 - 2$

9. $f(x) = -(x - 2)^4 + 8$

10. $f(x) = x^3 - 3x^2 + x - 1$

11. $f(x) = -x^3 + 3x^2 - 2x$

12. $f(x) = x^4 + 2x^3 + 4$

Name _____

What you should learn :

| 9.3 | How to factor polynomials and solve polynomial equations by factoring. |

Correlation to Pupil's Textbook:

Mid-Chapter Test (p. 487)   Chapter Test (p. 513)
Exercises 7–12   Exercises 5–8

Examples   *Factoring Polynomials and Solving Polynomial Equations*

a. $16x^3 - 80x^2 + 100x = 4x(4x^2 - 20x + 25)$ *Common monomial*

$\qquad\qquad\qquad\quad = 4x(2x - 5)^2$ *Perfect-square trinomial*

b.
$$3x^3 + 21x = 24x^2 \qquad \textit{Original equation}$$
$$3x^3 - 24x^2 + 21x = 0 \qquad \textit{Collect terms on the left side.}$$
$$3x(x^2 - 8x + 7) = 0 \qquad \textit{Factor out common monomial factor.}$$
$$3x(x - 7)(x - 1) = 0 \qquad \textit{Factor trinomial.}$$
$$3x = 0 \;\Rightarrow\; x = 0 \qquad \textit{Set first factor equal to 0.}$$
$$x - 7 = 0 \;\Rightarrow\; x = 7 \qquad \textit{Set second factor equal to 0.}$$
$$x - 1 = 0 \;\Rightarrow\; x = 1 \qquad \textit{Set third factor equal to 0.}$$

c.
$$x^3 + x^2 - 2x = 2 \qquad \textit{Original equation}$$
$$x^3 + x^2 - 2x - 2 = 0 \qquad \textit{Collect terms on the left side.}$$
$$(x^3 + x^2) - (2x + 2) = 0 \qquad \textit{Group terms (to factor by grouping).}$$
$$x^2(x + 1) - 2(x + 1) = 0 \qquad \textit{Factor out common monomial factors.}$$
$$(x + 1)(x^2 - 2) = 0 \qquad \textit{Factor out common binomial factor.}$$
$$x + 1 = 0 \;\Rightarrow\; x = -1 \qquad \textit{Set first factor equal to 0.}$$
$$x^2 - 2 = 0 \;\Rightarrow\; x = \pm\sqrt{2} \qquad \textit{Set second factor equal to 0.}$$

Guidelines: To solve a polynomial equation:
- Write the polynomial equation in standard form by collecting the terms on the left side of the equation.
- Factor the polynomial and apply the Zero-Product Property.

EXERCISES

In 1–6, factor completely with respect to the integers.

1. $27x^2 - 81x$ \qquad\qquad **2.** $8x^3 - 12x^2$ \qquad\qquad **3.** $x^3 - 2x^2 - 2x + 4$

4. $64x^3 - 27$ \qquad\qquad **5.** $20x^3 + 60x^2 + 45x$ \qquad\qquad **6.** $8x^3 + 27$

In 7–12, find all real-number solutions.

7. $3x^2 = 9x$ \qquad\qquad **8.** $x^2 = 2x + 15$ \qquad\qquad **9.** $9x^3 = 4x$

10. $x^2 + 2x = 24$ \qquad\qquad **11.** $x^3 + 6x^2 - x - 6 = 0$ \qquad\qquad **12.** $2x^3 - x^2 + 10x = 5$

Name _____

What you should learn :

9.4	How to divide polynomials using long division and synthetic division.

Correlation to Pupil's Textbook:

Mid-Chapter Test (p. 487)   **Chapter Test (p. 513)**
Exercises 13–18, 20   Exercises 11–13

Examples | Dividing Polynomials

a. Use long division to divide $x^4 - 3x^3 - 2x + 1$ by $x^2 + 1$.

$$
\begin{array}{r}
x^2 - 3x - 1 \\
x^2 + 1 \overline{)\, x^4 - 3x^3 \qquad - 2x + 1} \\
\underline{x^4 \qquad + x^2} \\
-3x^3 - x^2 - 2x \\
\underline{-3x^3 \qquad - 3x} \\
-x^2 + x + 1 \\
\underline{-x^2 \qquad - 1} \\
x + 2
\end{array}
$$

Save space for the "missing" terms.
Multiply x^2 times $x^2 + 1$.
Subtract.
Multiply $-3x$ times $x^2 + 1$.
Subtract.
Multiply -1 times $x^2 + 1$.
Subtract.

In fractional form, the result is $\dfrac{x^4 - 3x^3 - 2x + 1}{x^2 + 1} = x^2 - 3x - 1 + \dfrac{x+2}{x^2+1}.$

b. Use synthetic division to divide $x^3 + 3x^2 - 10x - 24$ by $x - 3$.
Then use the result to factor $x^3 + 3x^2 - 10x - 24$.

$$
\begin{array}{r|rrrr}
3 & 1 & 3 & -10 & -24 \\
  &   & 3 & 18 & 24 \\
\hline
  & 1 & 6 & 8   & 0
\end{array}
$$

The synthetic division array is shown at the right. From the array,
you can write the complete factorization as follows.

$$x^3 + 3x^2 - 10x - 24 = (x-3)(x^2+6x+8) = (x-3)(x+4)(x+2)$$

Guidelines:
To divide one polynomial by another polynomial:
- Use synthetic division when the divisor is of the form $x \pm k$. Otherwise, use long division.
- The Remainder Theorem (on page 482 of the textbook) can be used to evaluate $f(k)$.
- The Factor Theorem (on page 482 of the textbook) and synthetic division can be used to factor a polynomial.

EXERCISES

In 1 and 2, use long division. Write the result in fractional form.

1. $(2x^3 - x^2 - 3x + 4) \div (2x + 1)$

2. $(x^4 + 3x^3 - 3x^2 - 12x - 4) \div (x^2 + 3x + 1)$

In 3 and 4, use synthetic division.

3. $(x^3 - 6x^2 - 3x + 1) \div (x + 2)$

4. $(5x^4 - 2x^2 + 1) \div (x + 1)$

In 5 and 6, use the indicated zero to factor $f(x)$ completely.

5. $f(x) = 3x^3 - 11x^2 - 6x + 8, \quad 4$

6. $f(x) = x^3 - x^2 - 10x - 8, \quad -2$

Reteach
Chapter 9

Name _____

What you should learn :

9.5	How to find the rational zeros of a polynomial.

Correlation to Pupil's Textbook:

Chapter Test (p. 513)
Exercises 16, 17

Example   *Finding Rational Zeros of a Polynomial*

Find all real zeros of $f(x) = 3x^4 + x^3 - 8x^2 - 2x + 4$.

Begin by finding the rational zeros.

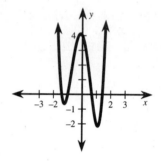

p: $\pm 1, \pm 2, \pm 4$ *Factors of constant term, 4*
q: $\pm 1, \pm 3$ *Factors of leading coefficient, 3*
$\frac{p}{q}$: $\pm 1, \pm 2, \pm 4, \pm \frac{1}{3}, \pm \frac{2}{3}, \pm \frac{4}{3}$ *List of possible rational zeros*

From the graph of the function at the right, you can see that
$x = -\frac{4}{3}$, $x = -1$, $x = \frac{2}{3}$, and $x = \frac{4}{3}$ are reasonable choices.
To test these choices, use synthetic division as follows.

$$-\frac{4}{3} \ \big| \ \begin{array}{ccccc} 3 & 1 & -8 & -2 & 4 \\ & -4 & 4 & \frac{16}{3} & -\frac{40}{9} \\ \hline 3 & -3 & -4 & \frac{10}{3} & -\frac{4}{9} \end{array}$$

Because the remainder of $-\frac{4}{9}$ is not zero, $-\frac{4}{3}$ is not a zero of the polynomial.

$$-1 \ \big| \ \begin{array}{ccccc} 3 & 1 & -8 & -2 & 4 \\ & -3 & 2 & 6 & -4 \\ \hline 3 & -2 & -6 & 4 & 0 \end{array}$$

Because the remainder is 0, -1 is a zero of the polynomial. The polynomial factors as
$f(x) = (x + 1)(3x^3 - 2x^2 - 6x + 4)$.

$$\frac{2}{3} \ \big| \ \begin{array}{cccc} 3 & -2 & -6 & 4 \\ & 2 & 0 & -4 \\ \hline 3 & 0 & -6 & 0 \end{array}$$

Because the remainder is zero, $\frac{2}{3}$ is a zero of the polynomial. The polynomial now factors as
$f(x) = (x + 1)(x - \frac{2}{3})(3x^2 - 6)$.

By setting each of the factors equal to zero, you can conclude that the
zeros of the polynomial are -1, $\frac{2}{3}$, $-\sqrt{2}$, and $\sqrt{2}$.

Guidelines: To find the rational zeros of a polynomial:
- Make a list of the possible rational zeros, using the Rational-Zero Test as stated on page 488 of the textbook.
- Sketch the graph of the function.
- Use the graph to select reasonable choices from the list and test each of these using synthetic division.

EXERCISES

In 1–8, find all real zeros of the polynomial function.

1. $f(x) = x^3 - 7x + 6$ **2.** $f(x) = x^3 + x^2 - 10x + 8$ **3.** $f(x) = 2x^3 - 3x^2 - 8x - 3$

4. $f(x) = x^3 + 3x^2 - 5x - 15$ **5.** $f(x) = x^3 - 5x^2 + 5x - 1$ **6.** $f(x) = x^3 - x^2 - 3x - 1$

7. $f(x) = x^3 + 8x^2 + 17x + 6$ **8.** $f(x) = 2x^4 + 7x^3 + 4x^2 - 7x - 6$

Reteach
Chapter 9

Name _____

What you should learn :

Correlation to Pupil's Textbook:

Chapter Test (p. 513)
Exercises 9, 10, 14, 15, 19, 20

| **Example** | *Writing a Polynomial as a Product of Linear Factors* |

Write $f(x) = x^4 - x^3 - 5x^2 - x - 6$ as a product of linear factors.

Because the polynomial is of degree 4, there are exactly four zeros (counting repeated and complex zeros). Begin by finding the rational zeros.

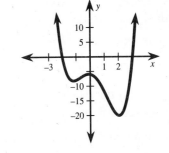

p: $\pm1, \pm2, \pm3, \pm6$ *Factors of constant term, -6*
q: ±1 *Factors of leading coefficient, 1*
$\frac{p}{q}$: $\pm1, \pm2, \pm3, \pm6$ *List of possible rational zeros*

From the graph of the function at the right, you can see that $x = -2$ and $x = 3$ are reasonable choices. To test these choices, use synthetic division as follows.

$$-2 \,\big|\, \begin{array}{ccccc} 1 & -1 & -5 & -1 & -6 \\ & -2 & 6 & -2 & 6 \\ \hline 1 & -3 & 1 & -3 & 0 \end{array}$$

Because the remainder is 0, -2 is a zero of the polynomial. The polynomial factors as
$f(x) = (x + 2)(x^3 - 3x^2 + x - 3)$.

$$3 \,\big|\, \begin{array}{cccc} 1 & -3 & 1 & -3 \\ & 3 & 0 & 3 \\ \hline 1 & 0 & 1 & 0 \end{array}$$

Because the remainder is zero, 3 is a zero of the polynomial. The polynomial now factors as
$f(x) = (x + 2)(x - 3)(x^2 + 1)$.

Because $(x^2 + 1)$ factors as $(x - i)(x + i)$, you can conclude that the complete (linear) factorization of the polynomial is

$$f(x) = (x + 2)(x - 3)(x - i)(x + i).$$

Guidelines: To write a polynomial function as a product of linear factors:
- Use the Fundamental Theorem of Algebra (on page 495 of the textbook) to determine the number of zeros.
- List all possible rational zeros.
- Sketch the graph of the function to determine the number of real zeros and the reasonable rational zeros to test.
- Use synthetic division to verify the zeros and to factor the polynomial.

EXERCISES

In 1–6, write the polynomial as a product of linear factors.

1. $x^4 - 6x^3 + 14x^2 - 54x + 45$

2. $x^4 - 5x^3 + 7x^2 - 5x + 6$

3. $x^3 + x^2 - 6x - 6$

4. $x^3 - 4x^2 + 4x - 16$

5. $x^4 + 3x^3 - 2x^2 + 6x - 8$

6. $x^4 - 16$

Name _____

What you should learn :

9.7	How to find the range and standard deviation of a collection of numbers.

Correlation to Pupil's Textbook:

Chapter Test (p. 513)
Exercise 18

Example	*Finding the Range and Standard Deviation*

The prices of seven brands of four-head VCR's with approximately the same features are as follows.

$239, $249, $259, $275, $299, $299, $319

Find the range and standard deviation of these prices.

The range is the difference of the highest and lowest prices. That is, the range is $319 - 239 = 80$.

To find the standard deviation, you must first find the mean, $\overline{x}$.

$$\overline{x} = \frac{239 + 249 + 259 + 275 + 2(299) + 319}{7} = 277$$

Now, you can calculate the standard deviation as follows.

$$s = \sqrt{\frac{239^2 + 249^2 + 259^2 + 275^2 + 2(299)^2 + 319^2}{7} - (277)^2} \qquad \textit{Use alternative formula.}$$

$$= \sqrt{\frac{542{,}391}{7} - 76{,}729} \qquad \textit{Simplify.}$$

$$\approx \sqrt{77{,}484.4286 - 76{,}729} \qquad \textit{Simplify.}$$

$$= \sqrt{755.4286} \qquad \textit{Simplify.}$$

$$\approx 27.49 \qquad \textit{Use a calculator.}$$

Thus, the standard deviation is approximately $27.49.

Guidelines: To find the standard deviation of a collection of numbers:
- Find the mean of the collection.
- Use a calculator to evaluate the alternative formula for standard deviation found on page 503 of the textbook.

EXERCISES

In 1 and 2, find the range and standard deviation of the data.

1. The weights (in pounds) of eleven children are as follows.

39, 52, 40, 45, 46, 55, 48, 40, 43, 47, 44

2. The grade point averages of twenty students are as follows.

1.1, 1.2, 1.2, 1.4, 1.4, 1.8, 1.8, 1.8, 1.8, 1.9
2.0, 2.2 2.4, 2.4, 2.4, 2.4, 2.4, 3.2 3.4, 3.6

Name _____

What you should learn:

| 10.1 | How to graph rational functions using vertical and horizontal asymptotes. |

Correlation to Pupil's Textbook:

Mid-Chapter Test (p. 540)   **Chapter Test (p. 567)**
Exercises 1–6, 20   Exercises 1, 2, 17

Examples   *Graphing Rational Functions*

a. Find the horizontal and vertical asymptotes of the graph of

$$g(x) = \frac{x}{x^2 + 4}.$$

Because the degree of the numerator is less than the degree of the denominator, the graph has $y = 0$ as a horizontal asymptote. Because there are no real x-values that make the denominator zero, the graph has no vertical asymptotes.

b. Find the horizontal and vertical asymptotes of the graph of

$$h(x) = \frac{x^4 + 1}{x^2 + x}.$$

Because the degree of the numerator is greater than the degree of the denominator, the graph has no horizontal asymptotes. Because the denominator is zero when x is 0 and -1, the graph has two vertical asymptotes: the lines $x = 0$ and $x = -1$.

c. Sketch the graph of $f(x) = \dfrac{1 - x^2}{x^2}$.

The graph has a horizontal asymptote of $y = -1$ and a vertical asymptote of $x = 0$.

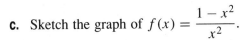

x	−2	−1	−0.5	0.5	1	2
y	−0.75	0	3	3	0	−0.75

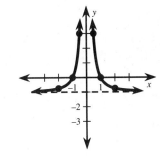

Plot the points in the table. Connect the points to the left of $x = 0$ with a smooth curve, and connect the points to the right of $x = 0$ with a different smooth curve. (Do not connect the two branches.) The graph is shown at the right.

Guidelines: To sketch the graph of a rational function:
- Find the vertical and horizontal asymptotes using the guidelines on page 519 of the textbook.
- Construct a table of values, choosing x-values to the left and right of any x-value that is a vertical asymptote.
- Plot the points and connect each branch with a smooth curve.

EXERCISES

In 1–4, identify all horizontal and vertical asymptotes.

1. $f(x) = \dfrac{x - 1}{x - 3}$ **2.** $f(x) = \dfrac{x}{(x + 1)^2}$ **3.** $f(x) = \dfrac{3x^2}{x^2 + 9}$ **4.** $f(x) = \dfrac{x^2 + 1}{x - 2}$

In 5–8, sketch the graph of the function.

5. $f(x) = \dfrac{2}{x + 1}$ **6.** $f(x) = \dfrac{3}{x^2 + 1}$ **7.** $f(x) = \dfrac{x - 1}{x + 2}$ **8.** $f(x) = \dfrac{x}{x^2 - 4}$

Name _____

What you should learn :

10.2	How to create models with inverse variation and with joint variation.

Examples | *Using Inverse and Joint Variation*

a. S varies inversely with t. When $t = \frac{3}{2}$, $S = 12$. Find the value of S when $t = 0.8$.

$$St = k \qquad\qquad \textit{Model for inverse variation}$$

$$(12)\left(\tfrac{3}{2}\right) = k \qquad \textit{Substitute 12 for S and } \tfrac{3}{2} \textit{ for } t.$$

$$18 = k \qquad\qquad \textit{Simplify.}$$

$$St = 18 \qquad\qquad \textit{Substitute 18 for k in the original model.}$$

$$S = \frac{18}{t} \qquad\qquad \textit{Divide both sides by } t.$$

$$S = \frac{18}{0.8} = 22.5 \qquad \textit{Substitute 0.8 for t and simplify.}$$

b. A varies jointly with x and the square of y. When $x = 10$ and $y = 9$, $A = 135$. Find the value of A when $x = 45$ and $y = 8$.

$$A = kxy^2 \qquad\qquad \textit{Model for joint variation}$$

$$135 = k(10)(9)^2 \qquad \textit{Substitute 135 for A, 10 for x, and 9 for y.}$$

$$135 = 810k \qquad\qquad \textit{Simplify.}$$

$$\tfrac{1}{6} = k \qquad\qquad \textit{Divide both sides by 810 and simplify.}$$

$$A = \tfrac{1}{6}xy^2 \qquad\qquad \textit{Substitute } \tfrac{1}{6} \textit{ for k in original model.}$$

$$A = \tfrac{1}{6}(45)(8)^2 = 480 \qquad \textit{Substitute 45 for x and 8 for y.}$$

Guidelines: To create and use models with inverse or joint variation:
- Choose the model for inverse or joint variation.
- Find the constant of variation.
- Write an equation that relates the variables.
- Evaluate the model to find the desired quantity.

EXERCISES

In 1–3, the variables x and y vary inversely. Use the given pair of values to find an equation that relates the variables. Find the value of y when $x = 2$.

1. $x = 10$, $y = \frac{1}{2}$ **2.** $x = -3$, $y = 3$ **3.** $x = \frac{1}{18}$, $y = 72$

In 4–6, the variable z varies jointly with the product of x and y. Use the given values to find an equation that relates the variables.

4. $x = 4$, $y = 3$, $z = 24$ **5.** $x = 8$, $y = -54$, $z = 144$ **6.** $x = 1$, $y = \frac{1}{8}$, $z = 4$

Reteach
Chapter 10

What you should learn :

| 10.3 | How to multiply and divide rational expressions and simplify the result. |

Correlation to Pupil's Textbook:

Mid-Chapter Test (p. 540) **Chapter Test (p. 567)**
Exercises 7–16   Exercises 5, 6

Examples | *Multiplying and Dividing Rational Expressions*

a. $\dfrac{x^2 - 2x}{x^2 + 2x + 1} \cdot \dfrac{x^2 + 4x + 3}{x^2 + 3x} = \dfrac{x(x-2)}{(x+1)^2} \cdot \dfrac{(x+3)(x+1)}{x(x+3)}$ *Factor each polynomial.*

$= \dfrac{x(x-2)(x+3)(x+1)}{(x+1)^2(x)(x+3)}$ *Multiply numerators and denominators.*

$= \dfrac{\cancel{x}(x-2)\cancel{(x+3)}\cancel{(x+1)}}{(x+1)^2\cancel{(x)}\cancel{(x+3)}}$ *Divide out common factors.*

$= \dfrac{x-2}{x+1}$ *Simplified form*

b. $\dfrac{x^3 - 8}{64x} \div \dfrac{x^2 - x - 2}{16x^2} = \dfrac{x^3 - 8}{64x} \cdot \dfrac{16x^2}{x^2 - x - 2}$ *Multiply by reciprocal.*

$= \dfrac{(x-2)(x^2 + 2x + 4)}{4(16x)} \cdot \dfrac{(16x)(x)}{(x-2)(x+1)}$ *Factor.*

$= \dfrac{(x-2)(x^2 + 2x + 4)(16x)(x)}{4(16x)(x-2)(x+1)}$ *Multiply.*

$= \dfrac{\cancel{(x-2)}(x^2 + 2x + 4)\cancel{(16x)}(x)}{4\cancel{(16x)}\cancel{(x-2)}(x+1)}$ *Divide out common factors.*

$= \dfrac{x(x^2 + 2x + 4)}{4(x+1)}$ *Simplified form*

Guidelines: To multiply or divide rational expressions:
- If dividing, multiply first expression by reciprocal of the second expression.
- Factor numerators and denominators.
- Multiply numerators and denominators.
- Divide out common factors to simplify.

EXERCISES

In 1–12, perform the indicated operation and simplify.

1. $\dfrac{12x^2 y}{5y^2} \cdot \dfrac{2xy}{3x^2}$

2. $\dfrac{4y^2}{9x} \cdot \dfrac{27}{16xy^2}$

3. $\dfrac{x^2 - 2x}{x^2 + 2x + 1} \cdot \dfrac{x^2 + 4x + 3}{x^2 + 3x}$

4. $\dfrac{x^2 + 2x - 3}{x + 2} \cdot \dfrac{x^2 + 2x}{x^2 - 1}$

5. $\dfrac{5t^5}{8} \div \dfrac{15t^2}{12}$

6. $\dfrac{48x^2}{y} \div \dfrac{36xy^2}{5}$

7. $\dfrac{8x^3 + 27}{2x^2 + 3x} \div \dfrac{4x^2 - 6x + 9}{3x^3}$

8. $\dfrac{x^2}{x^2 - 1} \div \dfrac{3x}{x + 1}$

9. $\dfrac{2x^3 - 12x^2}{x^2 - 4x - 12} \div \dfrac{8x^3 + 24x^2}{x^2 + 9x + 18}$

Reteach
Chapter 10

Name _____

What you should learn :

10.4	How to solve equations that contain rational expressions.

Correlation to Pupil's Textbook:

Chapter Test (p. 567)
Exercises 11–15, 18

Example	*Solving Rational Equations*

$$\frac{5x}{x-1} - 3 = \frac{2x+5}{x^2-1}$$ *Original equation*

$$\frac{5x}{x-1} - 3 = \frac{2x+5}{(x-1)(x+1)}$$ *Factor denominator.*

$$5x(x+1) - 3(x^2-1) = 2x+5$$ *Multiply both sides by least common denominator and simplify.*

$$5x^2 + 5x - 3x^2 + 3 = 2x+5$$ *Distribute.*

$$2x^2 + 3x - 2 = 0$$ *Write quadratic equation in standard form.*

$$(2x-1)(x+2) = 0$$ *Factor.*

$$x = \tfrac{1}{2}, \ -2$$ *Apply Zero-Product Property*

The solutions are $\frac{1}{2}$ and -2. Check these in the original equation.

Guidelines: To solve an equation that contains rational expressions:
- Find the least common denominator.
- Multiply *each* term on both sides of the equation by the least common denominator.
- Simplify and solve the resulting polynomial equation.

EXERCISES

In 1–12, solve the equation. Check each solution.

1. $\dfrac{3}{x} - \dfrac{2}{x+1} = \dfrac{4}{x}$

2. $\dfrac{1}{x-2} + \dfrac{1}{x+2} = \dfrac{4}{x^2-4}$

3. $\dfrac{2x}{x+3} + 5 = \dfrac{3}{x+3}$

4. $\dfrac{4}{x} - \dfrac{1}{x+2} = \dfrac{2}{x}$

5. $\dfrac{1}{x-2} + \dfrac{1}{x+3} = \dfrac{5}{x^2+x-6}$

6. $\dfrac{3}{x-1} - 6 = \dfrac{5x}{x-1}$

7. $\dfrac{5}{x} = \dfrac{2}{x+3} - \dfrac{3}{x}$

8. $\dfrac{1}{x-2} + \dfrac{1}{x+3} = \dfrac{17}{x^2+x-6}$

9. $\dfrac{3}{x-2} = 5 - \dfrac{2}{x-2}$

10. $\dfrac{5x}{x-1} - 2 = \dfrac{14}{x^2-1}$

11. $\dfrac{3}{x-8} - \dfrac{4}{x-2} = \dfrac{28}{x^2-10x+16}$

12. $\dfrac{2x}{x-2} - \dfrac{4x-1}{3x+2} = \dfrac{17x+4}{3x^2-4x-4}$

Scrambled answers for the first column of exercises: $-\frac{12}{7}$, -2, no solution, -4, $-\frac{1}{3}$, 3

Name _____

What you should learn :

10.5	How to add and subtract rational expressions and simplify complex fractions.

Examples *Working with Rational Expressions*

a. $\dfrac{5}{3x-12}+\dfrac{3x+1}{x^2-x-12}-\dfrac{2}{3}$ *Original expression*

$=\dfrac{5}{3(x-4)}+\dfrac{3x+1}{(x-4)(x+3)}-\dfrac{2}{3}$ *Factor denominators.*

$=\dfrac{5(x+3)}{3(x-4)(x+3)}+\dfrac{3(3x+1)}{3(x-4)(x+3)}-\dfrac{2(x-4)(x+3)}{3(x-4)(x+3)}$ *Rewrite fractions with LCD.*

$=\dfrac{5x+15}{3(x-4)(x+3)}+\dfrac{9x+3}{3(x-4)(x+3)}-\dfrac{2x^2-2x-24}{3(x-4)(x+3)}$ *Expand numerators.*

$=\dfrac{(5x+15)+(9x+3)-(2x^2-2x-24)}{3(x-4)(x+3)}$ *Add and subtract fractions.*

$=\dfrac{5x+15+9x+3-2x^2+2x+24}{3(x-4)(x+3)}$ *Distribute.*

$=\dfrac{-2x^2+16x+42}{3(x-4)(x+3)}$ *Simplify.*

b. $\dfrac{\left(\dfrac{6}{x-1}-3\right)}{\left(\dfrac{3}{x}\right)}=\dfrac{6(x)-3x(x-1)}{3(x-1)}=\dfrac{6x-3x^2+3x}{3(x-1)}=\dfrac{9x-3x^2}{3(x-1)}=\dfrac{3x-x^2}{x-1}$

Guidelines: To add or subtract rational expressions:
- Find the least common denominator (LCD).
- Rewrite each rational expression using the LCD.
- Add or subtract the numerators.

To simplify a complex fraction:
- Multiply numerator and denominator by the LCD of every fraction.

EXERCISES

In 1–9, perform the operations and simplify.

1. $\dfrac{5}{x}+\dfrac{2}{3x^2}$ **2.** $\dfrac{1}{2}+\dfrac{2}{x}-\dfrac{3}{x^2}$ **3.** $\dfrac{3}{x+5}-\dfrac{4}{x+1}$

4. $\dfrac{4x}{x^2-4}-\dfrac{3}{x+2}$ **5.** $\dfrac{2x-1}{x^2-x-2}-\dfrac{1}{x-2}$ **6.** $\dfrac{x}{x-1}+\dfrac{3x}{x^2-1}$

7. $\dfrac{\left(\dfrac{x^2}{x^2-1}\right)}{\left(\dfrac{3x}{x+1}\right)}$ **8.** $\dfrac{\left(2-\dfrac{1}{x}\right)}{x}$ **9.** $\dfrac{\left(1+\dfrac{1}{x}\right)}{\left(1-\dfrac{1}{x}\right)}$

Name _____

What you should learn :

10.6	How to find the monthly payment for an installment loan and calculate the total interest.

Correlation to Pupil's Textbook:

Chapter Test (p. 567)
Exercises 19, 20

Example	*Finding a Monthly Payment and Total Interest*

You are considering a home mortgage for $80,000 with an annual interest rate of 9% and monthly installments. Calculate the monthly payment and the total interest for a 15-year term.

$$M = P \cdot \frac{i}{1 - \left(\dfrac{1}{1+i}\right)^{12t}}$$ *Formula for monthly payment*

$$= 80,000 \cdot \frac{0.0075}{1 - \left(\dfrac{1}{1+0.0075}\right)^{12(15)}}$$ *Substitute 80,000 for P, 15 for t, and $\frac{0.09}{12} = 0.0075$ for i.*

$$\approx 811.41$$ *Use a calculator.*

The monthly payment is $811.41. To find the total interest paid over the 15-year term, subtract the loan amount ($80,000) from the total amount paid.

$$\boxed{\begin{array}{c}\text{Total} \\ \text{interest}\end{array}} = \boxed{\begin{array}{c}\text{Number of} \\ \text{payments}\end{array}} \cdot \boxed{\begin{array}{c}\text{Monthly} \\ \text{payment}\end{array}} - \boxed{\begin{array}{c}\text{Amount} \\ \text{of loan}\end{array}}$$

$$= (12)(15)(811.41) - 80,000$$

$$= 146,053.80 - 80,000$$

$$= 66,053.80$$

The total interest is $66,053.80.

Guidelines:
- To find the monthly payment for an installment loan, use the formula on page 556 of the textbook.
- To find the total interest paid, subtract the amount of the loan from the total amount paid over the term of the loan.

EXERCISES

In 1–3, find the monthly payment, the total amount paid, and the total interest paid.

1. A loan for $10,000 with an annual interest rate of 12% is taken out for a 3-year term.

2. A loan for $80,000 with an annual interest rate of 9% is taken out for a 30-year term.

3. A loan for $1000 with an annual interest rate of 16.5% is taken out for a 1-year term.

Reteach
Chapter 11

Name _____

What you should learn :

| 11.1 | How to sketch parabolas and write equations of parabolas. |

Correlation to Pupil's Textbook:

Mid-Chapter Test (p. 589)   **Chapter Test (p. 619)**
Exercises 1–6, 15, 20   Exercises 1, 13, 19

Examples — *Sketching Parabolas and Writing Equations of Parabolas*

a. Sketch the parabola given by $x^2 = 12y$.

Begin by writing the equation in standard form.

$$x^2 = 4py \quad \Rightarrow \quad 4p = 12 \quad \Rightarrow \quad p = 3$$

Because $p > 0$, the parabola opens upward. The focus is $(0,\ p) = (0,\ 3)$. The directrix is $y = -p$ or $y = -3$.

x	-2	-1	0	1	2
y	$\frac{1}{3}$	$\frac{1}{12}$	0	$\frac{1}{12}$	$\frac{1}{3}$

Plot the points, and connect them as shown at the right.

b. Write the standard form of the equation of the parabola with vertex at the origin and focus at $(-\frac{1}{4},\ 0)$. Sketch the parabola.

Because the focus is $(-\frac{1}{4},\ 0) = (p,\ 0)$, it follows that $p = -\frac{1}{4}$. The axis of the parabola is horizontal because it is the line that passes through the vertex and focus of the parabola.

$$y^2 = 4px \qquad \textit{Standard form (horizontal axis)}$$
$$y^2 = 4(-\tfrac{1}{4})x \qquad \textit{Substitute } -\tfrac{1}{4} \textit{ for } p.$$
$$y^2 = -x \qquad \textit{Standard form}$$

The graph is shown at the right.

Guidelines: To sketch a parabola or write its equation:
- If the x-variable is squared, the parabola has a vertical axis and the standard form is $x^2 = 4py$.
- If the y-variable is squared, the parabola has a horizontal axis and the standard form is $y^2 = 4px$.

EXERCISES

In 1–8, write the standard form of the equation of the parabola with vertex at the origin and the given focus.

1. $(0,\ \frac{1}{4})$ **2.** $(0,\ 1)$ **3.** $(-\frac{1}{2},\ 0)$ **4.** $(0,\ -3)$

5. $(-2,\ 0)$ **6.** $(\frac{1}{2},\ 0)$ **7.** $(0,\ -1)$ **8.** $(4,\ 0)$

In 9–12, sketch the parabola. Find the focus and directrix.

9. $x^2 = 2y$ **10.** $y^2 = 16x$ **11.** $4x^2 + y = 0$ **12.** $4x + y^2 = 0$

Name _____

What you should learn :

11.2	How to write an equation of a circle and find points of intersection.

Examples | *Writing Equations of Circles and Finding Points of Intersection*

a. Write the standard form of the equation of the circle that passes through the point $(2, -1)$ and whose center is the origin.

The radius of the circle is $r = \sqrt{(2-0)^2 + (-1-0)^2} = \sqrt{5}$.

$$x^2 + y^2 = r^2 \qquad \text{\textit{Standard form of the equation of a circle}}$$
$$x^2 + y^2 = (\sqrt{5})^2 \qquad \text{\textit{Substitute } } \sqrt{5} \text{ \textit{for} } r.$$
$$x^2 + y^2 = 5 \qquad \text{\textit{Simplify.}}$$

b. Find the points of intersection of the graphs of $x^2 = 2y$ and $-4x + 2y = 5$.

$$y = 2x + \tfrac{5}{2} \qquad \text{\textit{Solve the linear equation for } } y.$$
$$x^2 = 2y \qquad \text{\textit{Given equation of parabola}}$$
$$x^2 = 2(2x + \tfrac{5}{2}) \qquad \text{\textit{Substitute } } 2x + \tfrac{5}{2} \text{ \textit{for} } y.$$
$$x^2 = 4x + 5 \qquad \text{\textit{Simplify.}}$$
$$x^2 - 4x - 5 = 0 \qquad \text{\textit{Write quadratic equation in standard form.}}$$
$$(x-5)(x+1) = 0 \qquad \text{\textit{Factor.}}$$
$$x = 5, -1 \qquad \text{\textit{Zero-Product Property}}$$

When $x = 5$, $y = 2(5) + \tfrac{5}{2} = \tfrac{25}{2}$ and when $x = -1$, $y = 2(-1) + \tfrac{5}{2} = \tfrac{1}{2}$.
Thus, the points of intersection are $(5, \tfrac{25}{2})$ and $(-1, \tfrac{1}{2})$.

Guidelines:
- To write the equation of a circle with center at the origin, find the value of the radius and substitute into the equation $x^2 + y^2 = r^2$.
- To find the points of intersection of two graphs, use the method of substitution.

EXERCISES

In 1–4, write the standard form of the equation of the circle that passes through the point and whose center is the origin.

1. $(-3, 4)$ **2.** $(1, -9)$ **3.** $(2, 3)$ **4.** $(-1, -4)$

In 5–8, find the points of intersection, if any, of the graphs.

5. $x^2 + y^2 = 4$
   $x = y$

6. $x^2 + y^2 = 16$
   $x - 2y = -4$

7. $x^2 = 2y$
   $x + y = 4$

8. $4y^2 = x$
   $x + 4y = 3$

Reteach

Chapter 11

Name _____

What you should learn :

| **11.3** | How to write an equation of an ellipse, sketch an ellipse, and find its eccentricity. |

Correlation to Pupil's Textbook:

Mid-Chapter Test (p. 589)   **Chapter Test (p. 619)**

Exercises 9, 10, 13, 14, 16, 18, 19   Exercises 3, 15, 18

Examples   *Writing Equations of Ellipses and Sketching Ellipses*

a. Write an equation of the ellipse whose center is the origin and has a vertex at $(0, 5)$ and co-vertex at $(2, 0)$.

Because $(0, 5)$ is a vertex, you know that $a = 5$. Because $(2, 0)$ is a co-vertex, you know that $b = 2$. Because the major axis of the ellipse is vertical, its equation is

$$\frac{x^2}{b^2} + \frac{y^2}{a^2} = 1 \quad \text{or} \quad \frac{x^2}{4} + \frac{y^2}{25} = 1.$$

b. Find the foci of the ellipse in part **a**.

The foci are on the major axis, c units from the center. Using the equation

$$c^2 = a^2 - b^2 = 25 - 4 = 21$$

you can determine that $c = \sqrt{21}$. Thus the foci are

$$(0, \ c) = (0, \ \sqrt{21}) \quad \text{and} \quad (0, \ -c) = (0, \ -\sqrt{21}).$$

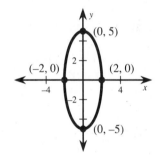

c. To sketch the ellipse in part **a**, plot the vertices, $(0, 5)$ and $(0, -5)$, and co-vertices, $(2, 0)$ and $(-2, 0)$. Then connect them with an oval-shaped curve as shown at the right.

d. The eccentricity of the ellipse in part **a** is

$$\text{eccentricity} = \frac{c}{a} = \frac{\sqrt{21}}{5} \approx 0.9165.$$

Guidelines: To sketch an ellipse:
- The line segment joining the vertices is the major axis.
- Plot the vertices and co-vertices and connect them with an oval-shaped curve.

EXERCISES

In 1–4, write the standard form of the equation of the ellipse. (The center of the ellipse is the origin.)

1. Vertex: $(3, 0)$
Co-vertex: $(0, 1)$

2. Vertex: $(0, -4)$
Co-vertex: $(3, 0)$

3. Vertex: $(-4, 0)$
Co-vertex: $(0, 2)$

4. Vertex: $(0, 5)$
Co-vertex: $(-2, 0)$

In 5–8, sketch the ellipse and find its eccentricity.

5. $\dfrac{x^2}{4} + \dfrac{y^2}{9} = 1$

6. $\dfrac{x^2}{8} + \dfrac{y^2}{4} = 1$

7. $\dfrac{x^2}{9} + y^2 = 1$

8. $\dfrac{x^2}{9} + \dfrac{y^2}{34} = 1$

© D.C. Heath and Company

Name _____

What you should learn :

11.4	How to write an equation of a hyperbola and sketch a hyperbola.

Correlation to Pupil's Textbook:

Chapter Test (p. 619)
Exercises 4, 16, 20

Examples	*Writing Equations of Hyperbolas and Sketching Hyperbolas*

a. Write an equation of the hyperbola whose foci are $(4, 0)$ and $(-4, 0)$, whose vertices are $(3, 0)$ and $(-3, 0)$, and whose center is the origin.

Because the line passing through the vertices is horizontal, the transverse axis is horizontal. Thus, the standard form of the equation of the hyperbola is as shown at the right. The distance between each vertex and the origin is $a = 3$. The distance between each foci and the origin is $c = 4$. Using the equation $b^2 = c^2 - a^2 = 16 - 9 = 7$, you can determine that $b = \sqrt{7}$.

$$\frac{x^2}{a^2} - \frac{y^2}{b^2} = 1$$

$$\frac{x^2}{9} - \frac{y^2}{7} = 1$$

b. Sketch the hyperbola given by $\dfrac{y^2}{4} - \dfrac{x^2}{9} = 1$.

Because the y^2-term is positive, the transverse axis is vertical, and the standard form of the equation is

$$\frac{y^2}{a^2} - \frac{x^2}{b^2} = 1$$

where $a = 2$ and $b = 3$. The vertices are $(0, a) = (0, 2)$ and $(0, -a) = (0, -2)$. Plot these points. Sketch a rectangle that is centered at the origin, $2a = 4$ units high and $2b = 6$ units wide. To sketch the asymptotes, draw lines connecting the opposite corners of the rectangle. Sketch the hyperbola above and below the rectangle—be sure to include the vertices.

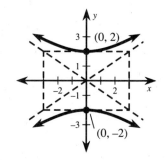

Guidelines: To sketch a hyperbola whose equation is in standard form:
- If the x^2-term is positive, the transverse axis is horizontal and the vertices are $(a, 0)$ and $(-a, 0)$. If the y^2-term is positive, the transverse axis is vertical and the vertices are $(0, a)$ and $(0, -a)$.
- Follow the technique used in Example **b** above.

EXERCISES

In 1–3, write the standard form of the equation of the hyperbola. (The center is the origin.)

1. Foci: $(-5, 0), (5, 0)$
Vertices: $(-2, 0), (2, 0)$

2. Foci: $(-3, 0), (3, 0)$
Vertices: $(-2, 0), (2, 0)$

3. Foci: $(0, -7), (0, 7)$
Vertices: $(0, -1), (0, 1)$

In 4–7, sketch the hyperbola.

4. $x^2 - 2y^2 = 2$

5. $\dfrac{y^2}{9} - \dfrac{x^2}{16} = 1$

6. $\dfrac{x^2}{4} - \dfrac{y^2}{9} = 1$

7. $\dfrac{y^2}{25} - \dfrac{x^2}{9} = 1$

Reteach
Chapter 11

Name _____

What you should learn :

11.5	How to write equations of conics and sketch conics that have been translated.

Correlation to Pupil's Textbook:

Chapter Test (p. 619)
Exercises 9–12

Examples	*Sketching and Writing Equations of Translated Conics*

a. Write an equation of the ellipse whose vertices are $(2, \ 1)$ and $(2, \ -5)$ and whose co-vertices are $(0, \ -2)$ and $(4, \ -2)$.

The major axis is vertical. The center is mid-way between the vertices: $(h, \ k) = (2, \ -2)$. The distance between the center and a vertex is $a = 3$. The distance between the center and a co-vertex is $b = 2$.

$$\frac{(x-h)^2}{b^2} + \frac{(y-k)^2}{a^2} = 1 \qquad \textit{Standard form of translated ellipse with a vertical major axis}$$

$$\frac{(x-2)^2}{4} + \frac{(y+2)^2}{9} = 1 \qquad \textit{Substitute 2 for h, -2 for k, 2 for b, and 3 for a.}$$

b. Write the equation $y = -\frac{1}{4}x^2 + 2x - 5$ in standard form and sketch its graph.

$$
\begin{aligned}
y &= -\tfrac{1}{4}x^2 + 2x - 5 & &\textit{Original equation} \\
4y &= -x^2 + 8x - 20 & &\textit{Multiply both sides by 4.} \\
x^2 - 8x &= -4y - 20 & &\textit{Group terms.} \\
x^2 - 8x + 16 &= -4y - 20 + 16 & &\textit{Complete the square.} \\
(x-4)^2 &= -4y - 4 & &\textit{Simplify.} \\
(x-4)^2 &= -4(y+1) & &\textit{Standard form} \\
(x-h)^2 &= 4p(y-k) & &\textit{Standard form}
\end{aligned}
$$

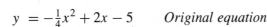

From the standard form, you can see that the vertex is $(h, \ k) = (4, \ -1)$. Because $p = -1$ is negative, the parabola opens downward.

Guidelines: To sketch the graph of a translated conic:
- Complete the square for each second-degree variable.
- Write the equation in standard form.
- Follow the techniques outlined in Lessons 11.1–11.4.

EXERCISES

In 1–4, write the equation in standard form. Then sketch the graph.

1. Parabola: $y^2 - 2y - 4x - 7 = 0$

2. Hyperbola: $-9x^2 + 16y^2 + 54x + 64y - 161 = 0$

3. Circle: $x^2 + y^2 + 2x + 6y + 6 = 0$

4. Ellipse: $4x^2 + y^2 - 8x - 4y + 4 = 0$

Reteach
Chapter 11

Name _____

What you should learn :

11.6	How to classify a conic and how to solve quadratic systems of equations.

Correlation to Pupil's Textbook:

Chapter Test (p. 619)
Exercises 5–8, 17

Examples | *Classifying Conics and Solving Quadratic Systems*

a. *Equation* $B^2 - 4AC$ *Conic*

$5x^2 - 3y^2 + 2x + 3y - 4 = 0$ $0^2 - 4(5)(-3) = 60 > 0$ Hyperbola

$4x^2 + 4xy + 6y^2 - 2x + 3y = 0$ $4^2 - 4(4)(6) = -80 < 0$ Ellipse $(A \neq C)$

$2x^2 + 2x + 2y - 5 = 0$ $0^2 - 4(2)(0) = 0$ Parabola

$2x^2 + 2y^2 + 2x - 5 = 0$ $0^2 - 4(2)(2) = -16 < 0$ Circle $(A = C)$

b. Find the points of intersection of the graphs of the system.

$$\begin{cases} x^2 + y^2 - 8y + 7 = 0 & \text{Equation 1} \\ -x^2 + y - 1 = 0 & \text{Equation 2} \end{cases}$$

Because the second equation has no y^2-term, solve that equation for y and use the substitution technique.

$$y = x^2 + 1 \quad \text{Solve for } y \text{ in Equation 2.}$$
$$x^2 + (x^2 + 1)^2 - 8(x^2 + 1) + 7 = 0 \quad \text{Substitute } x^2 + 1 \text{ for } y \text{ in Equation 1.}$$
$$x^2 + x^4 + 2x^2 + 1 - 8x^2 - 8 + 7 = 0 \quad \text{Expand terms.}$$
$$x^4 - 5x^2 = 0 \quad \text{Collect like terms.}$$
$$x^2(x^2 - 5) = 0 \quad \text{Factor.}$$
$$x = 0, \pm\sqrt{5} \quad \text{Zero-Product Property}$$

When $x = 0$, $y = 0^2 + 1 = 1$. When $x = \sqrt{5}$, $y = (\sqrt{5})^2 + 1 = 6$. When $x = -\sqrt{5}$, $y = (-\sqrt{5})^2 + 1 = 6$. Thus, the points of intersection are $(0, 1)$, $(\sqrt{5}, 6)$, and $(-\sqrt{5}, 6)$.

Guidelines:
- To classify the equation of a conic, find the value of the discriminant, $B^2 - 4AC$, and use the table on page 606 of the textbook.
- To find the points of intersection of a quadratic system, use the substitution or elimination techniques.

EXERCISES

In 1–5, classify the conic.

1. $4x^2 - 4y^2 - 2x - 4y - 5 = 0$ **2.** $3y^2 + 2x + 3y - 1 = 0$ **3.** $x^2 - 2x + 3y - 5 = 0$

4. $x^2 + 3y^2 - x + 2y - 4 = 0$ **5.** $3x^2 + 3y^2 + 3x + 3y - 1 = 0$

In 6–8, find the points of intersection of the graphs of the equations.

6. $x^2 - y - 2 = 0$ **7.** $x^2 + y^2 - 3 = 0$ **8.** $x^2 + y^2 + 4x - 4y + 4 = 0$

    $2x^2 + y - 6x - 7 = 0$ $2x^2 - y = 0$ $x^2 + y^2 - 4x - 4y + 4 = 0$

Name _____

What you should learn :

12.1	How to write sequences, use sigma notation, and evaluate a series.

Correlation to Pupil's Textbook:

Mid-Chapter Test (p. 644)   **Chapter Test (p. 671)**
Exercises 4–6   Exercises 3, 6

Examples | *Working with Sequences and Series*

a. Write the first five terms of the sequence whose nth term is $a_n = -4n+3$. Assume that n begins with 1.

$$a_1 = -4(1) + 3 = -1$$
$$a_2 = -4(2) + 3 = -5$$
$$a_3 = -4(3) + 3 = -9$$
$$a_4 = -4(4) + 3 = -13$$
$$a_5 = -4(5) + 3 = -17$$

b. Write the first five terms of the sequence whose nth term is $a_n = (-1)^{n+1}(\frac{1}{n})$. Assume that n begins with 1.

$$a_1 = (-1)^{1+1}(\tfrac{1}{1}) = 1$$
$$a_2 = (-1)^{2+1}(\tfrac{1}{2}) = -\tfrac{1}{2}$$
$$a_3 = (-1)^{3+1}(\tfrac{1}{3}) = \tfrac{1}{3}$$
$$a_4 = (-1)^{4+1}(\tfrac{1}{4}) = -\tfrac{1}{4}$$
$$a_5 = (-1)^{5+1}(\tfrac{1}{5}) = \tfrac{1}{5}$$

c. Evaluate the series $\sum\limits_{k=0}^{6} k!$.

$$\sum_{k=0}^{6} k! = 0! + 1! + 2! + 3! + 4! + 5! + 6!$$
$$= 1 + 1 + 2 + 6 + 24 + 120 + 720$$
$$= 874$$

d. Evaluate the series $\sum\limits_{j=1}^{4} (1-j)$.

$$\sum_{j=1}^{4} (1-j) = (1-1) + (1-2) + (1-3) + (1-4)$$
$$= 0 + (-1) + (-2) + (-3)$$
$$= -6$$

Guidelines: To write sequences and series:
- The domain of a sequence is the set of positive integers.
- The sum of the first n terms of a sequence is called a series and is represented by

$$\sum_{i=1}^{n} a_i = a_1 + a_2 + a_3 + \cdots + a_n.$$

EXERCISES

In 1–8, write the first five terms of the sequence. Begin with $n = 1$.

1. $a_n = 5n - 3$

2. $a_n = n - 2$

3. $a_n = 3(2)^n$

4. $a_n = \dfrac{(-1)^n}{n}$

5. $a_n = \dfrac{n!}{(n+2)!}$

6. $a_n = (\tfrac{1}{3})^n$

7. $a_n = \dfrac{n}{n^2 + 1}$

8. $a_n = 3 + (-1)^n$

In 9–12, evaluate the series.

9. $\sum\limits_{i=1}^{4} (3i + 1)$

10. $\sum\limits_{j=0}^{2} 3(-\tfrac{1}{2})^j$

11. $\sum\limits_{k=2}^{6} (-1)^k (2k)$

12. $\sum\limits_{n=1}^{5} (n^2 + n)$

Name _____

What you should learn :

12.2	How to find the nth term and the sum of an arithmetic sequence.

Correlation to Pupil's Textbook:

Mid-Chapter Test (p. 644)   **Chapter Test (p. 671)**
Exercises 1, 2, 7, 9–12,   Exercises 1, 4, 5, 7
16, 18, 19, 21

Examples	*Working with Arithmetic Sequences*

a. Write a formula for the nth term of the arithmetic sequence.

$2, 11, 20, 29, 38, \ldots$

The common difference of the sequence is $d = 11 - 2 = 9$.

$a_n = a_1 + (n-1)d$	*Formula for the nth term of an arithmetic sequence.*
$a_n = 2 + (n-1)9$	*Substitute 2 for a_1 and 9 for d.*
$a_n = 9n - 7$	*Simplify.*

b. The twelfth term of an arithmetic sequence is -26. The twentieth term is -50. Write a formula for the nth term.

The twentieth term is 8 times the common difference, d, plus the twelfth term.

$$a_{20} = a_{12} + 8d \quad \Rightarrow \quad -50 = -26 + 8d \quad \Rightarrow \quad d = -3$$

The twelfth term is 11 times the common difference, d, plus the first term.

$$a_{12} = a_1 + 11d \quad \Rightarrow \quad -26 = a_1 + 11(-3) \quad \Rightarrow \quad a_1 = 7$$

The formula for the nth term is

$$a_n = a_1 + (n-1)d = 7 + (n-1)(-3) = -3n + 10$$

c. The sum of the first 18 terms of the arithmetic sequence whose nth term is $a_n = 3n - 1$ (where n begins with 1) is

$$\text{sum} = n\left(\frac{a_1 + a_n}{2}\right) = 18\left(\frac{2 + 53}{2}\right) = 495.$$

Guidelines:

- The nth term of an arithmetic sequence is $a_n = a_1 + (n-1)d$, where a_1 is the first term and d is the common difference.

- The sum of the first n terms of an arithmetic sequence is $n\left(\dfrac{a_1 + a_n}{2}\right)$.

EXERCISES

In 1–4, find a formula for the nth term of the arithmetic sequence.

1. $a_1 = 5, d = -4$ **2.** $a_3 = 15, d = -2$ **3.** $a_2 = 12, d = 3$ **4.** $a_2 = 15, d = \frac{3}{2}$

In 5–7, find the indicated term of the arithmetic sequence.

5. $a_1 = 4, d = 10$, 9th term   **6.** $a_1 = 5, d = 8$, 8th term   **7.** $a_1 = 2, d = 7$, 17th term

In 8–10, find the sum of the first n terms of the arithmetic sequence.

8. $2, 3, 4, 5, 6, \ldots; n = 19$ **9.** $25, 35, 45, 55, 65, \ldots; n = 50$ **10.** $3, 6, 9, 12, 15, \ldots; n = 50$

Name _____

What you should learn :

12.3	How to find the nth term and sum of a finite geometric sequence.

Correlation to Pupil's Textbook:

Mid-Chapter Test (p. 644)   Chapter Test (p. 671)
Exercises 3, 8, 13–15, 17, 20   Exercises 2, 8, 25

Examples | *Working with Geometric Sequences*

a. Find the sum of the first ten terms of the geometric sequence.

$$-2, \ 3, \ -\tfrac{9}{2}, \ \tfrac{27}{4}, \ -\tfrac{81}{8}, \ \ldots$$

The common ratio of the sequence is $r = \frac{3}{-2} = -\frac{3}{2}$. Because $a_1 = -2$,
the nth term of the sequence is $a_n = a_1(r)^{n-1} = -2(-\frac{3}{2})^{n-1}$.

$$\sum_{i=1}^{10} -2(-\tfrac{3}{2})^{i-1} = a_1\left(\frac{1-r^n}{1-r}\right) \qquad \textit{Formula for the sum of a geometric sequence}$$

$$= -2\left(\frac{1-(-\tfrac{3}{2})^{10}}{1-(-\tfrac{3}{2})}\right) \qquad \textit{Substitute } -2 \textit{ for } a_1, -\tfrac{3}{2} \textit{ for } r, \textit{ and } 10 \textit{ for } n.$$

$$\approx 45.332 \qquad \textit{Use a calculator.}$$

b. The third term of a geometric sequence is 8 and the sixth term is -64.
Write a formula for the nth.

The sixth term can be obtained by multiplying the third term by r^3. Solve
for r as follows.

$$a_3 r^3 = a_6 \quad \Rightarrow \quad 8r^3 = -64 \quad \Rightarrow \quad r^3 = -8 \quad \Rightarrow \quad r = -2$$

Solve for the first term as follows.

$$a_1 r^{3-1} = a_3 \quad \Rightarrow \quad a_1(-2)^2 = 8 \quad \Rightarrow \quad a_1 = 2$$

The formula for the nth term is $a_n = a_1 r^{n-1} = 2(-2)^{n-1}$

Guidelines:
- The nth term of a geometric sequence is $a_1 r^{n-1}$, where a_1 is the first term and r is the common ratio.
- The sum of the first n terms of a geometric sequence is $a_1\left(\frac{1-r^n}{1-r}\right)$.

EXERCISES

In 1–4, find a formula for the n-th term of the geometric sequence.

1. $a_1 = 2, r = -\frac{1}{3}$ **2.** $a_1 = 5, r = 1.1$ **3.** $a_1 = 3, r = -1$ **4.** $a_1 = 8, r = \frac{1}{2}$

In 5–7, find the indicated term of the geometric sequence.

5. $a_1 = 4, r = \frac{1}{2}$, 7th term   **6.** $a_1 = 5, r = 1.1$, 20th term   **7.** $a_1 = 2, a_4 = -16$, 6th term

In 8–10, find the sum of the first n terms of the geometric sequence.

8. $4, 2, 1, \frac{1}{2}, \ldots; n = 10$ **9.** $1, \frac{2}{3}, \frac{4}{9}, \frac{8}{27}, \ldots; n = 8$ **10.** $3, 0.9, 0.27, 0.081, \ldots; n = 5$

Name _____

What you should learn :

| 12.4 | How to find the sum of an infinite geometric series (if it has one). |

Correlation to Pupil's Textbook:

Chapter Test (p. 671)
Exercises 9, 10, 26

Examples | *Summing an Infinite Geometric Series*

a. Find the sum. $\displaystyle\sum_{n=0}^{\infty} 2(0.1)^n$

The series has a sum because $|r| = 0.1 < 1$.

$\displaystyle\sum_{n=0}^{\infty} 2(0.1)^n = 2 + 0.2 + 0.002 + 0.0002 + \cdots$

Expand the series to identify $a_1 = 2$.

$\displaystyle\sum_{n=0}^{\infty} a_1(r)^n = \frac{a_1}{1-r}$

Formula for the sum of an infinite geometric series

$= \dfrac{2}{1 - 0.1}$

Substitute 2 for a_1 and 0.1 for r.

$= \dfrac{20}{9}$

Simplify.

b. Find the sum. $\displaystyle\sum_{n=1}^{\infty} 7(-\tfrac{3}{4})^{n-1}$

The series has a sum because $|r| = \left|-\tfrac{3}{4}\right| = \tfrac{3}{4} < 1$.

$\displaystyle\sum_{n=1}^{\infty} 7(-\tfrac{3}{4})^{n-1} = 7 + 7(-\tfrac{3}{4}) + 7(-\tfrac{3}{4})^2 + \cdots$

Expand the series to identify $a_1 = 7$.

$\displaystyle\sum_{n=1}^{\infty} a_1(r)^{n-1} = \frac{a_1}{1-r}$

Formula for the sum of an infinite geometric series

$= \dfrac{7}{1 - (-\tfrac{3}{4})}$

Substitute 7 for a_1 and $-\tfrac{3}{4}$ for r.

$= 4$

Simplify.

Guidelines: To find the sum of an infinite geometric series:
- The infinite series has a sum if $|r| < 1$. It does not have a sum if $|r| \geq 1$.
- If it has a sum, it is $\dfrac{a_1}{1-r}$, where a_1 is the first term and r is the common ratio.

EXERCISES

In 1–12, find the sum (if it has one).

1. $\displaystyle\sum_{n=0}^{\infty} 3(\tfrac{1}{2})^n$

2. $\displaystyle\sum_{n=0}^{\infty} 3(\tfrac{4}{3})^n$

3. $\displaystyle\sum_{n=0}^{\infty} 2(-\tfrac{1}{2})^n$

4. $\displaystyle\sum_{n=1}^{\infty} 5(\tfrac{1}{10})^{n-1}$

5. $\displaystyle\sum_{n=0}^{\infty} (0.9)^n$

6. $\displaystyle\sum_{n=1}^{\infty} 9(0.7)^{n-1}$

7. $\displaystyle\sum_{n=0}^{\infty} -7(\tfrac{1}{3})^n$

8. $\displaystyle\sum_{n=0}^{\infty} 2(\tfrac{5}{4})^n$

9. $\displaystyle\sum_{n=0}^{\infty} (-0.4)^n$

10. $\displaystyle\sum_{n=0}^{\infty} 2$

11. $\displaystyle\sum_{n=1}^{\infty} 4(-\tfrac{1}{3})^{n-1}$

12. $\displaystyle\sum_{n=0}^{\infty} 0.6(0.1)^n$

Name _____

What you should learn:

12.5	How to evaluate binomial coefficients and expand a binomial using the Binomial Theorem.

Examples | *Using the Binomial Theorem*

a. $\dbinom{10}{7} = \dfrac{10!}{7!3!} = \dfrac{10 \cdot 9 \cdot 8 \cdot 7!}{7! \cdot 3!} = \dfrac{10 \cdot 9 \cdot 8}{3 \cdot 2 \cdot 1} = 120$

b. $\dbinom{4}{4} = \dfrac{4!}{4!0!} = \dfrac{1}{0!} = 1$

c. $\dbinom{6}{5} = \dfrac{6!}{5!1!} = \dfrac{6 \cdot 5!}{5!1!} = \dfrac{6}{1} = 6$

d. $(t+2)^4 = \dbinom{4}{0}t^4 + \dbinom{4}{1}t^3(2) + \dbinom{4}{2}t^2(2^2) + \dbinom{4}{3}t(2^3) + \dbinom{4}{4}(2^4)$ *Binomial Theorem*

$= (1)t^4 + (4)t^3(2) + (6)t^2(2^2) + (4)t(2^3) + (1)(2^4)$ *Evaluate binomial coefficients.*

$= (1)t^4 + (4)t^3(2) + (6)t^2(4) + (4)t(8) + (1)(16)$ *Evaluate powers.*

$= t^4 + 8t^3 + 24t^2 + 32t + 16$ *Simplify.*

e. $(5-3x)^3 = \dbinom{3}{0}5^3 + \dbinom{3}{1}5^2(-3x) + \dbinom{3}{2}5(-3x)^2 + \dbinom{3}{3}(-3x)^3$ *Binomial Theorem*

$= (1)5^3 + (3)5^2(-3x) + (3)5(-3x)^2 + (1)(-3x)^3$ *Evaluate binomial coefficients.*

$= (1)(125) + (3)(25)(-3x) + (3)(5)(9x^2) + (1)(-27x^3)$ *Evaluate powers.*

$= 125 - 225x + 135x^2 - 27x^3$ *Simplify.*

Guidelines: To expand a binomial using the Binomial Theorem:

- Use the formula on page 651 of the textbook.
- The coefficient of $x^{n-m}y^m$ is $\dbinom{n}{m} = \dfrac{n!}{m!(n-m)!}$.

EXERCISES

In 1–9, expand the binomial.

1. $(x-3)^4$ **2.** $(t+4)^3$ **3.** $(2x+1)^5$

4. $(3-2x)^3$ **5.** $(2a-b)^3$ **6.** $(3x+y)^3$

7. $(1-x)^6$ **8.** $(x-3y)^4$ **9.** $(x+2)^7$

Name _____

| 12.6 | How to find the balance and the amount of monthly deposit in an increasing annuity. |

Correlation to Pupil's Textbook:

Chapter Test (p. 671)
Exercises 23, 24

Examples | *Finding the Balance and Amount of Deposit in an Annuity*

a. $200 is deposited in an increasing annuity at the end of each month. The annual interest rate is $6\frac{1}{2}\%$, compounded monthly. Find the balance at the end of one year.

$$A = D\left[\frac{(1+i)^{nt} - 1}{i}\right]$$ *Formula for balance in an increasing annuity.*

$$= 200\left[\frac{\left(1 + \frac{0.065}{12}\right)^{(12)(1)} - 1}{\frac{0.065}{12}}\right]$$ *Substitute 200 for D, 12 for n, 1 for t, and $\frac{0.065}{12}$ for i.*

$$\approx 2472.81$$ *Use a calculator.*

The balance at the end of one year is $2472.81.

b. Find the amount of deposit necessary each month to produce a balance of $50,000 for a child's college education. The annual interest rate is 6%, compounded monthly. Assume an 18-year period.

$$D = A\left[\frac{i}{(1+i)^{nt} - 1}\right]$$ *Formula for the amount of deposit in an increasing annuity.*

$$= 50,000\left[\frac{\frac{0.06}{12}}{\left(1 + \frac{0.06}{12}\right)^{(12)(18)} - 1}\right]$$ *Substitute 50,000 for A, 12 for n, 18 for t, and $\frac{0.06}{12}$ for i.*

$$\approx 129.08$$ *Use a calculator.*

The monthly deposit should be $129.08

Guidelines:
- To find the balance of an increasing annuity, use the formula on page 659 of the textbook.
- To find the amount of deposit necessary to attain a specified balance in an annuity, use the formula on page 660 of the textbook.

EXERCISES

1. $20 is deposited in an increasing annuity at the end of each month. The annual interest rate is 6%, compounded monthly. What is the balance at the end of five years?

2. $100 is deposited in an increasing annuity at the end of each month. The annual interest rate is 8%, compounded monthly. What is the balance at the end of forty years? Find the total amount deposited.

3. Find the amount of deposit necessary each month to produce a balance of $500,000 in 20 years, if the annual interest rate is 8%, compounded monthly.

Name _____

What you should learn :

13.1	How to evaluate the trigonometric functions of an acute angle and use trigonometric identities.

Correlation to Pupil's Textbook:

Mid-Chapter Test (p. 702)   **Chapter Test (p. 727)**
Exercises 1–3, 20, 21   Exercise 19

Examples *Using Trigonometric Relationships*

a. Find the six trigonometric functions of the acute angle θ, given that $\sin\theta = \frac{2}{5}$.

Begin by constructing a right triangle as shown at the right. Because $\sin\theta = \frac{\text{opp.}}{\text{hyp.}}$, you can assign the value of 2 to the side opposite θ and the value of 5 to the hypotenuse. Label the third side a.

$$a^2 + b^2 = c^2 \qquad \textit{Pythagorean Theorem}$$
$$a^2 + 4 = 25 \qquad \textit{Substitute 2 for b and 5 for c.}$$
$$a = \sqrt{21} \qquad \textit{Solve for a.}$$

The six trigonometric functions of θ are as follows.

$$\sin\theta = \frac{\text{opp.}}{\text{hyp.}} = \frac{2}{5} \qquad \cos\theta = \frac{\text{adj.}}{\text{hyp.}} = \frac{\sqrt{21}}{5} \qquad \tan\theta = \frac{\text{opp.}}{\text{adj.}} = \frac{2}{\sqrt{21}}$$
$$\csc\theta = \frac{\text{hyp.}}{\text{opp.}} = \frac{5}{2} \qquad \sec\theta = \frac{\text{hyp.}}{\text{adj.}} = \frac{5}{\sqrt{21}} \qquad \cot\theta = \frac{\text{adj.}}{\text{opp.}} = \frac{\sqrt{21}}{2}$$

b. $\sec\theta = \dfrac{1}{\cos\theta}$ If $\cos\theta = \frac{2}{3}$, then $\sec\theta = \frac{3}{2}$.

$\sin\theta = \dfrac{1}{\csc\theta}$ If $\csc\theta = \frac{5}{2}$, then $\sin\theta = \frac{2}{5}$.

$\tan\theta = \dfrac{1}{\cot\theta}$ If $\cot\theta = \sqrt{2}$, then $\tan\theta = \frac{1}{\sqrt{2}} = \frac{\sqrt{2}}{2}$.

c. Find $\sin\theta$, given that $\cos\theta = \frac{1}{4}$.

$$\sin^2\theta + \cos^2\theta = 1 \qquad \textit{Pythagorean Identity}$$
$$\sin^2\theta + (\tfrac{1}{4})^2 = 1 \qquad \textit{Substitute } \tfrac{1}{4} \textit{ for } \cos\theta.$$
$$\sin^2\theta = \tfrac{15}{16} \qquad \textit{Subtract } \tfrac{1}{16} \textit{ from both sides.}$$
$$\sin\theta = \tfrac{\sqrt{15}}{4} \qquad \textit{Take the square root of both sides.}$$

Guidelines:
- Know the right-triangle definitions on page 678 of the textbook.
- Know the trigonometric function values of the common angles 30°, 45°, and 60°, given on page 679 of the textbook.
- Know the fundamental identities given on page 680 of the textbook.

EXERCISES

In 1–8, sketch a right triangle that has θ as one of its acute angles. Then find the values of the five trigonometric functions that are not given.

1. $\cos\theta = \frac{12}{13}$ **2.** $\sec\theta = \frac{4}{3}$ **3.** $\tan\theta = \frac{3}{4}$ **4.** $\sin\theta = \frac{1}{2}$

5. $\cot\theta = \frac{5}{2}$ **6.** $\csc\theta = 3$ **7.** $\sec\theta = \frac{6}{5}$ **8.** $\cos\theta = \frac{1}{4}$

Name _____

What you should learn :

13.2	How to measure angles in standard position using degree measure and radian measure.

Correlation to Pupil's Textbook:

Mid-Chapter Test (p. 702)   Chapter Test (p. 727)
Exercises 4–12   Exercise 1

Examples	*Measuring Angles in Standard Position*

a. Find two angles, one with positive measure and one with negative measure, that are coterminal with $\theta = 100°$.

$100° + 360° = 460°$ *These angles are coterminal with $\theta = 100°$ because they have*
$100° - 360° = -260°$ *the same initial and terminal sides as $\theta = 100°$.*

b. Find the complement and the supplement of $\theta = \frac{2\pi}{7}$.

Complement: $\frac{\pi}{2} - \frac{2\pi}{7} = \frac{3\pi}{14}$ *The sum of $\frac{2\pi}{7}$ and $\frac{3\pi}{14}$ is $\frac{\pi}{2}$.*

Supplement: $\pi - \frac{2\pi}{7} = \frac{5\pi}{7}$ *The sum of $\frac{2\pi}{7}$ and $\frac{5\pi}{7}$ is π.*

c. Rewrite the degree measure as radians.

$$20° = (20 \text{ degrees})\left(\frac{\pi \text{ radians}}{180 \text{ degrees}}\right) = \frac{\pi}{9} \text{ radians}$$

d. Rewrite the radian measure as degrees.

$$-\frac{2\pi}{3} \text{ radians} = \left(-\frac{2\pi}{3} \text{ radians}\right)\left(\frac{180 \text{ degrees}}{\pi \text{ radians}}\right) = -\frac{360\pi}{3\pi} \text{ degrees} = -120°$$

Guidelines:

- Two angles are coterminal if they have the same initial and terminal sides.
- Two angles are complementary if the sum of their measures is $\frac{\pi}{2}$.
- Two angles are supplementary if the sum of their measures is π.
- To convert between degrees and radians, use the conversion factors given on page 687 of the textbook.

EXERCISES

In 1–4, find two angles, one with a positive measure and the other with a negative measure, that are coterminal with the given angle.

1. $105°$ **2.** $-75°$ **3.** $\frac{4\pi}{3}$ **4.** $-\frac{\pi}{4}$

In 5–8, find the complement and the supplement of the angle.

5. $14°$ **6.** $87°$ **7.** $\frac{\pi}{3}$ **8.** $\frac{4\pi}{9}$

In 9–12, rewrite the degree measure as radians.

9. $315°$ **10.** $-75°$ **11.** $2°$ **12.** $37°$

In 13–16, rewrite the radian measure as degrees.

13. $\frac{5\pi}{6}$ **14.** $-\frac{3\pi}{4}$ **15.** $\frac{2\pi}{9}$ **16.** 3

Reteach
Chapter 13

Name _____

What you should learn :

13.3	How to evaluate trigonometric functions of any angle.

Correlation to Pupil's Textbook:

Mid-Chapter Test (p. 702)   **Chapter Test (p. 727)**
Exercises 13–19, 22   Exercise 2

Examples   *Evaluating Trigonometric Functions*

a. Let $(3, -2)$ be a point on the terminal side of θ. Find $\sin\theta$, $\cos\theta$, and $\tan\theta$.

Because the point is $(x, y) = (3, -2)$, you know that $x = 3$ and $y = -2$.

$$\begin{aligned}
r &= \sqrt{x^2 + y^2} && \textit{Formula for r}\\
&= \sqrt{3^2 + (-2)^2} && \textit{Substitute 3 for x and } -2 \textit{ for y.}\\
&= \sqrt{13} && \textit{Simplify.}
\end{aligned}$$

The sine, cosine, and tangent of θ are as follows.

$$\sin\theta = \frac{y}{r} = \frac{-2}{\sqrt{13}} = -\frac{2}{\sqrt{13}}, \qquad \cos\theta = \frac{x}{r} = \frac{3}{\sqrt{13}}, \qquad \tan\theta = \frac{y}{x} = \frac{-2}{3} = -\frac{2}{3}$$

b. Find $\sec 225°$.

As shown at the right, $\theta = 225°$ lies in Quadrant III. The sign of $\sec\theta$ in Quadrant III is negative (it is the same as the sign of $\cos\theta$ in Quadrant III). The reference angle is $225° - 180° = 45°$.

$$\sec 225° = -\sec 45° = -\sqrt{2}$$

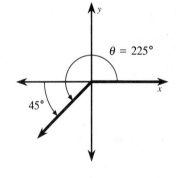

c. Find $\sin(-\frac{13\pi}{6})$.

$\theta = -\frac{13\pi}{6}$ lies in Quadrant IV. The sign of $\sin\theta$ in Quadrant IV is negative. The positive angle that is coterminal with $-\frac{13\pi}{6}$ is $\frac{11\pi}{6}$. The reference angle is $2\pi - \frac{11\pi}{6} = \frac{\pi}{6}$.

$$\sec(-\tfrac{13\pi}{6}) = -\sin\tfrac{\pi}{6} = -\tfrac{1}{2}$$

Guidelines:

To evaluate a trigonometric function of an angle θ:
- Determine the quadrant in which θ lies.
- Determine the sign of the trigonometric function of θ from the chart of quadrant signs on page 695 of the textbook.
- Find the reference angle θ'.
- Evaluate the trigonometric function for the angle θ'.

EXERCISES

In 1–12, evaluate the function without using a calculator.

1. $\tan 135°$ **2.** $\sin(-60°)$ **3.** $\cos 210°$ **4.** $\sec(-315°)$

5. $\cot \frac{7\pi}{6}$ **6.** $\csc \frac{2\pi}{3}$ **7.** $\tan \frac{7\pi}{3}$ **8.** $\sin(-\frac{3\pi}{4})$

9. $\cos \frac{15\pi}{4}$ **10.** $\csc \frac{5\pi}{6}$ **11.** $\cot \frac{11\pi}{6}$ **12.** $\sec(-\frac{4\pi}{3})$

Reteach
Chapter 13

Name _____

What you should learn :

| **13.4** | How to evaluate inverse trigonometric functions. |

Correlation to Pupil's Textbook:

Chapter Test (p. 727)
Exercises 4–6, 20

Examples *Evaluating Inverse Trigonometric Functions*

a. $\arctan(-1) = \theta$ *To evaluate* $\arctan(-1)$, *set it equal to* θ.

$\tan\theta = -1$ θ *is the angle whose tangent is* -1.

$\theta = -45°$ *For arctangent,* $-90° < \theta < 90°$; $\tan(-45°) = -1$

$\theta = -\frac{\pi}{4}$ $-45° = -\frac{\pi}{4}$ *radians.*

b. $\arcsin\frac{1}{2} = \theta$ *To evaluate* $\arcsin\frac{1}{2}$, *set it equal to* θ.

$\sin\theta = \frac{1}{2}$ θ *is the angle whose sine is* $\frac{1}{2}$.

$\theta = 30°$ *For arcsine,* $-90° \leq \theta \leq 90°$; $\sin 30° = \frac{1}{2}$

$\theta = \frac{\pi}{6}$ $30° = \frac{\pi}{6}$ *radians.*

c. $\arccos(-\frac{1}{2}) = \theta$ *To evaluate* $\arccos(-\frac{1}{2})$, *set it equal to* θ.

$\cos\theta = -\frac{1}{2}$ θ *is the angle whose cosine is* $-\frac{1}{2}$.

$\theta = 120°$ *For arccosine,* $0° \leq \theta \leq 180°$; $\cos 120° = -\frac{1}{2}$

$\theta = \frac{2\pi}{3}$ $120° = \frac{2\pi}{3}$ *radians.*

d. $\arcsin(-\frac{\sqrt{2}}{2}) = \theta$ *To evaluate* $\arcsin(-\frac{\sqrt{2}}{2})$, *set it equal to* θ.

$\sin\theta = -\frac{\sqrt{2}}{2}$ θ *is the angle whose sine is* $-\frac{\sqrt{2}}{2}$.

$\theta = -45°$ *For arcsine,* $-90° \leq \theta \leq 90°$; $\sin(-45°) = -\frac{\sqrt{2}}{2}$

$\theta = -\frac{\pi}{4}$ $-45° = -\frac{\pi}{4}$ *radians.*

Guidelines: To evaluate inverse trigonometric functions:
- Use the definitions on page 703 of the textbook.
- For arcsine, $-90° \leq \theta \leq 90°$.
- For arccosine, $0° \leq \theta \leq 180°$.
- For arctangent, $-90° < \theta < 90°$.

EXERCISES

In 1–12, evaluate the expression. Write the result in degrees and in radians.

1. $\arcsin\frac{\sqrt{3}}{2}$ **2.** $\arctan 1$ **3.** $\arccos\frac{1}{2}$ **4.** $\arctan(-\frac{1}{\sqrt{3}})$

5. $\arctan\sqrt{3}$ **6.** $\arcsin\frac{1}{\sqrt{2}}$ **7.** $\arccos(-\frac{\sqrt{3}}{2})$ **8.** $\arcsin(-\frac{1}{2})$

9. $\arccos(-\frac{\sqrt{2}}{2})$ **10.** $\arctan\frac{1}{\sqrt{3}}$ **11.** $\arcsin(-\frac{1}{\sqrt{2}})$ **12.** $\arccos(-\frac{1}{2})$

© D.C. Heath and Company

Name _____

What you should learn :

13.5	How to use the Law of Sines to solve a triangle.

Example	*Using the Law of Sines*

Solve the triangle in which $a = 11$, $b = 15$, and $A = 37°$.

$\dfrac{\sin B}{b} = \dfrac{\sin A}{a}$ *Law of Sines*

$\dfrac{\sin B}{15} = \dfrac{\sin 37°}{11}$ *Substitute 15 for b, 11 for a, and 37° for A.*

$\sin B = 15\left(\dfrac{\sin 37°}{11}\right)$ *Multiply both sides by 15.*

$\sin B \approx 0.821$ *Use a calculator (set in degree mode).*

$B \approx 55.2°$ *or* $124.8°$ *There are two angles between 0° and 180° whose sine is 0.821.*

Case 1

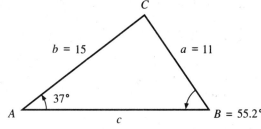

$B = 55.2°$

$C = 180° - 37° - 55.2° = 87.8°$

$\dfrac{c}{\sin C} = \dfrac{a}{\sin A}$

$c = (\sin 87.8°)\left(\dfrac{11}{\sin 37°}\right) \approx 18.3$

Solutions: $B \approx 55.2°$, $C \approx 87.8°$, $c \approx 18.3$

Case 2

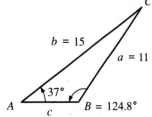

$B = 124.8°$

$C = 180° - 37° - 124.8° = 18.2°$

$\dfrac{c}{\sin C} = \dfrac{a}{\sin A}$

$c = (\sin 18.2°)\left(\dfrac{11}{\sin 37°}\right) \approx 5.7$

Solutions: $B \approx 124.8°$, $C \approx 18.2°$, $c \approx 5.7$

Guidelines: To solve an oblique triangle using the Law of Sines:
- If two angles and a side are given, there is a unique solution.
- If two sides and an angle are given, there may be no solution, a unique solution, or two solutions.

EXERCISES

In 1–6, solve the triangle.

1. $A = 23°$, $B = 57°$, $c = 12$ **2.** $A = 23°$, $B = 57°$, $a = 12$ **3.** $B = 34°$, $C = 108°$, $b = 20$

4. $A = 42°$, $a = 10$, $b = 21$ **5.** $B = 55°$, $a = 20$, $b = 25$ **6.** $A = 30°$, $a = 20$, $b = 38$

Name _____

What you should learn :

| **13.6** | How to use the Law of Cosines to solve a triangle. |

Correlation to Pupil's Textbook:

Chapter Test (p. 727)
Exercises 9, 10, 12–17

| **Examples** | *Using the Law of Cosines* |

a. The diagonals of a parallelogram have lengths of 22 centimeters and 36 centimeters. The diagonals intersect at an angle of 58° as shown at the right. Find the length of one of the shorter sides of the parallelogram.

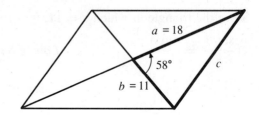

Because the lengths of the diagonals are 22 and 36, it follows that $a = 18$ and $b = 11$. To calculate c, the side opposite angle $C = 58°$, use the Law of Cosines.

$$c^2 = a^2 + b^2 - 2ab \cos C \qquad \textit{Law of Cosines}$$

$$c^2 = (18)^2 + (11)^2 - 2(18)(11) \cos 58° \qquad \textit{Substitute 18 for a, 11 for b, and 58° for C.}$$

$$c^2 \approx 253.1520 \qquad \textit{Use a calculator.}$$

$$c \approx 15.3 \qquad \textit{Take the square root of both sides.}$$

The length of one of the shorter sides is approximately 15.3 centimeters.

b. Find angle A in the triangle in which $a = 22$, $b = 19$, and $c = 14$.

$$\cos A = \frac{b^2 + c^2 - a^2}{2bc} \qquad \textit{Law of Cosines}$$

$$\cos A = \frac{(19)^2 + (14)^2 - (22)^2}{2(19)(14)} \qquad \textit{Substitute 19 for b, 14 for c, and 22 for a.}$$

$$A \approx 82.1° \qquad \textit{Use a calculator.}$$

Guidelines: To solve a triangle using the Law of Cosines:
- This law is appropriate for those triangles for which three sides are given or for which two sides and an included angle are given.
- Use the appropriate form from those listed on page 716 of the textbook.

EXERCISES

In 1–6, solve the triangle.

1. $A = 62°$, $b = 56$, $c = 40$

2. $B = 100°$, $a = 12$, $c = 13$

3. $C = 42°$, $a = 22$, $b = 35$

4. $a = 7$, $b = 7$, $c = 10$

5. $a = 39$, $b = 14$, $c = 27$

6. $a = 19$, $b = 21$, $c = 13$

Reteach
Chapter 14

Name _____

Examples | *Graphing the Sine and Cosine Functions*

a. Sketch one cycle of the graph of $y = 3 \sin 2x$.

The amplitude of the graph is 3, and the period of the graph is $\frac{2\pi}{|2|} = \pi$. Thus, the graph cycles once from 0 to π. To begin, divide the interval $[0, \ \pi]$ into four equal segments, and label each, as shown below.

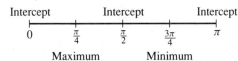

For the graph of $y = 3 \sin 2x$, the intercepts occur when $x = 0$, $x = \frac{\pi}{2}$, and $x = \pi$. The maximum occurs when $x = \frac{\pi}{4}$ and the minimum occurs when $x = \frac{3\pi}{4}$, as shown at the right.

b. Sketch one cycle of the graph of $y = 2 \cos \frac{\pi}{3}x$.

The amplitude of the graph is 2, and the period of the graph is

$$\frac{2\pi}{|\pi/3|} = 6.$$ Thus, the graph cycles once from 0 to 6. To begin, divide the interval $[0, \ 6]$ into four equal segments, and label each, as shown below.

For the graph of $y = 2 \cos \frac{\pi}{3}x$, the intercepts occur when $x = \frac{3}{2}$ and $x = \frac{9}{2}$. The maximum occurs when $x = 0$ and $x = 6$, and the minimum occurs when $x = 3$, as shown at the right.

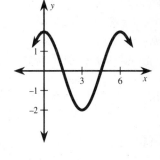

Guidelines: To sketch the graph of $y = a \sin bx$ or $y = a \cos bx$:

- The amplitude is $|a|$ and the period is $\dfrac{2\pi}{|b|}$.

- Divide the interval for one cycle into four equal segments.
- Plot the intercepts, maximum points, and minimum points.
- Connect the points with a smooth curve.

EXERCISES

In 1–8, sketch one cycle of the graph of the function.

1. $y = 4 \sin x$ **2.** $y = 2 \cos x$ **3.** $y = \sin 2x$ **4.** $y = \cos 3x$

5. $y = \sin \dfrac{x}{3}$ **6.** $y = 2 \sin 2x$ **7.** $y = 3 \sin \dfrac{x}{3}$ **8.** $y = 2 \cos 2\pi x$

What you should learn :

		Correlation to Pupil's Textbook:
14.2	How to graph shifts and reflections of the graphs of the sine and cosine functions.	**Mid-Chapter Test (p. 752)** **Chapter Test (p. 777)**
		Exercises 4–6, 16–20 Exercises 1–3, 20

Examples   *Graphing Shifts and Reflections of Sine and Cosine Graphs*

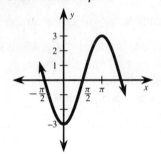

a. Sketch one cycle of the graph of $y = -3\sin(x + \frac{\pi}{2})$.

The amplitude of the graph is $|-3| = 3$. The period of the graph is $\frac{2\pi}{1} = 2\pi$. Because

$$\frac{c}{b} = \frac{\frac{\pi}{2}}{1} = \frac{\pi}{2} > 0$$

the graph is shifted $\frac{\pi}{2}$ units to the left. The graph cycles once from $-\frac{\pi}{2}$ to $\frac{3\pi}{2}$. Because $a = -3 < 0$, reflect the sine curve in the x-axis. The graph is shown at the right.

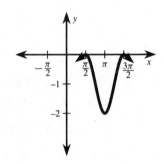

b. Sketch one cycle of the graph of $y = \cos(2x - \pi) - 1$.

The amplitude of the graph is 1. The period of the graph is $\frac{2\pi}{2} = \pi$. Because

$$\frac{c}{b} = \frac{-\pi}{2} < 0$$

the graph is shifted $\frac{\pi}{2}$ units to the right. The graph cycles once from $\frac{\pi}{2}$ to $\frac{3\pi}{2}$. Because $d = -1 < 0$, the graph is shifted down 1 unit. The graph is shown at the right.

Guidelines: To sketch the graph of $y = a\sin(bx + c) + d$ or $y = a\cos(bx + c) + d$:

- Follow the outline for shifts on page 737 of the textbook.
- If $a < 0$, reflect the graph in the line $y = d$.

EXERCISES

In 1–9, sketch one cycle of the graph of the function.

1. $y = 2 + \sin 2x$

2. $y = \cos x - 1$

3. $y = 4\sin x + 1$

4. $y = \sin(x + \pi)$

5. $y = 3\cos(x - \frac{\pi}{2})$

6. $y = 2\cos(x + \pi)$

7. $y = 1 + 2\sin(2x + \pi)$

8. $y = 4\cos(x + \frac{\pi}{2}) - 2$

9. $y = -2\cos(x + \frac{3\pi}{2})$

Name _____

What you should learn :

14.3 How to use and verify trigonometric identities.	

Correlation to Pupil's Textbook:

Mid-Chapter Test (p. 752) **Chapter Test (p. 777)**
Exercises 10–15   Exercises 4–6, 8

Examples *Using and Verifying Trigonometric Identities*

a. $\csc(-x) - \csc(-x)\cos^2 x = \csc(-x)(1 - \cos^2 x)$ *Factor out* $\csc(-x)$.

$$= \csc(-x)(\sin^2 x)$$ *Pythagorean identity*

$$= \frac{1}{\sin(-x)} \cdot \sin^2 x$$ *Reciprocal identity*

$$= \frac{1}{-\sin x} \cdot \sin^2 x$$ *Negative angle identity*

$$= -\sin x$$ *Simplify.*

b. $1 - 2\sin^2 x = 1 - 2(1 - \cos^2 x)$ *Pythagorean identity*

$$= 1 - 2 + 2\cos^2 x$$ *Distributive Property*

$$= 2\cos^2 x - 1$$ *Simplify.*

c. $\dfrac{\sin x + 1}{\cos x + \cot x} = \dfrac{\sin x + 1}{\cos x + \dfrac{\cos x}{\sin x}}$ *Cotangent identity*

$$= \frac{\sin x(\sin x + 1)}{\sin x \cos x + \cos x}$$ *Multiply numerator and denominator by* $\sin x$.

$$= \frac{\sin x(\sin x + 1)}{\cos x(\sin x + 1)}$$ *Factor.*

$$= \frac{\sin x}{\cos x}$$ *Divide by common factor.*

$$= \tan x$$ *Tangent identity*

Guidelines: To verify a trigonometric identity:

- Work on one side of the equation at a time.
- Look for ways to use the fundamental identities listed on pages 680 and 744 of the textbook.
- Look for ways to add fractions or factor an expression.

EXERCISES

In 1–8, verify the trigonometric identity.

1. $\tan\left(\frac{\pi}{2} - x\right)\sin x = \cos x$

2. $\dfrac{\tan^2 x}{\sec x} = \sec x - \cos x$

3. $\dfrac{\sin x}{1 + \cos x} = \dfrac{1 - \cos x}{\sin x}$

4. $\sec x \cot x = \csc x$

5. $\cos^2 x(1 + \tan^2 x) = 1$

6. $\sin\left(\frac{\pi}{2} - x\right)\sec x = 1$

7. $\sec(-x)\cot(-x)\sin(-x) = 1$

8. $\cos x(\csc x + \tan x) = \cot x + \sin x$

What you should learn :

14.4	How to solve a trigonometric equation.

Correlation to Pupil's Textbook:

Chapter Test (p. 777)
Exercises 10, 11, 13, 18, 19

Examples	*Solving Trigonometric Equations*

Each of the following equations are solved for $0 \leq x < 2\pi$.

a.

$\sqrt{2}\cos x \tan x = \tan x$	*Original equation*
$\sqrt{2}\cos x \tan x - \tan x = 0$	*Subtract $\tan x$ from both sides.*
$\tan x(\sqrt{2}\cos x - 1)\, 0$	*Factor.*
$\tan x = 0 \implies x = 0,\ \pi$	*Set first factor equal to 0.*
$\cos x = \frac{1}{\sqrt{2}} \implies x = \frac{\pi}{4},\ \frac{7\pi}{4}$	*Set second factor equal to 0.*

b.

$\cot^2 x - 1 = 2$	*Original equation*
$\cot^2 x = 3$	*Add 1 to both sides.*
$\cot x = \pm\sqrt{3}$	*Take square root of both sides.*
$x = \frac{\pi}{6},\ \frac{5\pi}{6},\ \frac{7\pi}{6},\ \frac{11\pi}{6}$	*Solve for x.*

c.

$\sin x - 2\cos^2 x + 1 = 0$	*Original equation*
$\sin x - 2(1 - \sin^2 x) + 1 = 0$	*Pythagorean identity*
$\sin x - 2 + 2\sin^2 x + 1 = 0$	*Distributive Property*
$2\sin^2 x + \sin x - 1 = 0$	*Write in standard quadratic form.*
$(2\sin x - 1)(\sin x + 1) = 0$	*Factor.*
$\sin x = \frac{1}{2} \implies x = \frac{\pi}{6},\ \frac{5\pi}{6}$	*Set first factor equal to 0.*
$\sin x = -1 \implies x = \frac{3\pi}{2}$	*Set second factor equal to 0.*

Guidelines: To solve a trigonometric equation:
- If the equation contains only one term with a trigonometric function, isolate that term and solve.
- If the equation contains more than one trigonometric term, move all terms to the left side of the equation and factor if possible.
- Look for ways to use the fundamental identities to write the equation with only one trigonometric function.

EXERCISES

In 1–12, solve the equation for $0 \leq x \leq 2\pi$.

1. $7\tan x + 9 = 2$ **2.** $\tan x \sec x = \sqrt{2}\tan x$ **3.** $\sqrt{3}\sin x = 2\sin x \cos x$

4. $\sin^3 x - \sin x = 0$ **5.** $2\cos^2 x = 1$ **6.** $2\sin^2 x + \sin x = 1$

Name _____

What you should learn :

| 14.5 | How to use sum and difference formulas. |

| Examples | *Using Sum and Difference Formulas* |

a. Find the exact value of $\tan 15°$.

$$\tan 15° = \tan(45° - 30°)$$ *Substitute $45° - 30°$ for $15°$.*

$$= \frac{\tan 45° - \tan 30°}{1 + \tan 45° \tan 30°}$$ *Formula for the tangent of a difference*

$$= \frac{1 - \frac{\sqrt{3}}{3}}{1 + (1)(\frac{\sqrt{3}}{3})}$$ *Substitute 1 for $\tan 45°$ and $\frac{\sqrt{3}}{3}$ for $\tan 30°$.*

$$= \frac{\left(1 - \frac{\sqrt{3}}{3}\right)(\sqrt{3})}{\left(1 + \frac{\sqrt{3}}{3}\right)(\sqrt{3})}$$ *Multiply the numerator and denominator by $\sqrt{3}$.*

$$= \frac{\sqrt{3} - 1}{\sqrt{3} + 1}$$ *Simplify.*

b. Find the exact value of $\cos \frac{13\pi}{12}$.

$$\cos \frac{13\pi}{12} = \cos\left(\frac{\pi}{4} + \frac{5\pi}{6}\right)$$ *Substitute $\frac{\pi}{4} + \frac{5\pi}{6}$ for $\frac{13\pi}{12}$.*

$$= \cos\frac{\pi}{4}\cos\frac{5\pi}{6} - \sin\frac{\pi}{4}\sin\frac{5\pi}{6}$$ *Formula for the cosine of a sum*

$$= \left(\frac{\sqrt{2}}{2}\right)\left(-\frac{\sqrt{3}}{2}\right) - \left(\frac{\sqrt{2}}{2}\right)\left(\frac{1}{2}\right)$$ *Substitute values for sine and cosine of $\frac{\pi}{4}$ and $\frac{5\pi}{6}$.*

$$= -\frac{\sqrt{6}}{4} - \frac{\sqrt{2}}{4}$$ *Multiply.*

$$= -\frac{\sqrt{6} + \sqrt{2}}{4}$$ *Add fractions.*

Guidelines: To find the exact value of a trigonometric function:
- Write the angle as the sum or difference of common angles.
- Expand using the appropriate formulas from page 760 of the textbook.

EXERCISES

In 1–8, find the exact value of the expression.

1. $\cos 15°$ **2.** $\sin 195°$ **3.** $\cos 75°$ **4.** $\sec 75°$

5. $\sin \frac{7\pi}{12}$ **6.** $\tan \frac{13\pi}{12}$ **7.** $\sin \frac{\pi}{12}$ **8.** $\cos \frac{11\pi}{12}$

Reteach

Chapter 14

Name ______________________________

What you should learn:

14.6 How to use double-angle formulas and half-angle formulas.

Correlation to Pupil's Textbook:

Chapter Test (p. 777)
Exercises 7, 9, 12, 21

Examples *Using Double-Angle and Half-Angle Formulas*

a.

$\dfrac{\cos 2x}{\cos x + \sin x} = 0$	*Original equation*
$\dfrac{\cos^2 x - \sin^2 x}{\cos x + \sin x} = 0$	*Use identity for* $\cos 2x$.
$\dfrac{(\cos x - \sin x)(\cos x + \sin x)}{\cos x + \sin x} = 0$	*Factor the numerator.*
$\cos x - \sin x = 0$	*Divide out common factor.*
$1 - \tan x = 0$	*Divide both sides by* $\cos x$.
$\tan x = 1$	*Isolate* $\tan x$.
$x = \frac{\pi}{4}, \frac{5\pi}{4}$	*Solve for* x.

b. Use a half-angle formula to evaluate $\sin(-\frac{7\pi}{8})$.

Begin by noting that $-\frac{7\pi}{8}$ lies in Quadrant III, and the sine is negative there.

$\sin(-\frac{7\pi}{8}) = -\sqrt{\dfrac{1 - \cos(-\frac{7\pi}{4})}{2}}$	*Use the formula for* $\sin \frac{u}{2}$ *with* $-\frac{7\pi}{8} = \frac{1}{2}(-\frac{7\pi}{4})$.
$= -\sqrt{\dfrac{1 - \frac{\sqrt{2}}{2}}{2}}$	*Substitute* $\frac{\sqrt{2}}{2}$ *for* $\cos(-\frac{7\pi}{4})$.
$= \sqrt{\dfrac{2 - \sqrt{2}}{4}}$	*Multiply the numerator and denominator by* 2.

Guidelines: To use double-angle and half-angle formulas:

- Become familiar with all three formulas for $\cos 2u$ and use the most convenient form in a given problem.
- The signs of $\sin \frac{u}{2}$ and $\cos \frac{u}{2}$ depend on the quadrant in which $\frac{u}{2}$ lies.

EXERCISES

In 1–4, solve the equation for $0 \le x < 2\pi$.

1. $\sin 2x + \sqrt{3} \cos x = 0$

2. $\cos 2x - \cos x - 2 = 0$

3. $\sin x = \cos 2x$

4. $2 \cos 2x + 4 \sin x - 3 = 0$

5. Find $\cos \frac{u}{2}$ if $\cos u = \frac{4}{5}$, $0 \le u < \frac{\pi}{2}$.

6. Find $\sin \frac{u}{2}$ if $\cos u = -\frac{2}{3}$, $\frac{\pi}{2} \le u < \pi$.

7. Find $\cos \frac{u}{2}$ if $\cos u = 0.1$, $\frac{3\pi}{2} \le u < 2\pi$.

8. Find $\tan \frac{u}{2}$ if $\sin u = \frac{1}{3}$, $0 \le u < \frac{\pi}{2}$.

Reteach
Chapter 15

Name _____

What you should learn :

| 15.1 | How to find the probability of an event. |

Correlation to Pupil's Textbook:

Mid-Chapter Test (p. 804)   Chapter Test (p. 828)
Exercises 1–3   Exercises 1, 13

Examples   *Finding the Probability of an Event*

a. Find the probability of choosing an E when selecting a letter at random from those in the word COLLEGE.

$$P = \frac{\text{Number of E's in the word COLLEGE}}{\text{Number of letters in the word COLLEGE}} = \frac{2}{7}$$

b. Thirty students in an *Algebra 2* class took a test: 3 received A's, 10 received B's, 10 received C's, 5 received D's, and 2 failed the test. If a student from the class is chosen at random, what is the probability that the student received a B on the test?

$$P = \frac{\text{Number who received B's}}{\text{Number who took the test}} = \frac{10}{30} = \frac{1}{3}$$

c. In a group of 10 children, 3 have blond hair, 2 have brown hair, and 5 have black hair. If a child is chosen at random from the group, what is the probability that he or she has brown hair?

$$P = \frac{\text{Number who have brown hair}}{\text{Number in the group}} = \frac{2}{10} = \frac{1}{5}$$

What is the probability that the chosen child has red hair?

$$P = \frac{\text{Number who have red hair}}{\text{Number in the group}} = \frac{0}{10} = 0$$

Guidelines: To find the probability of an event:
- The probability, P, that an event will occur is the ratio of the number of outcomes in the event to the number of outcomes in the sample space.

EXERCISES

1. A multiple choice test question has 5 possible choices. What is the probability of randomly selecting the correct answer?

2. What is the probability of choosing an A, B, or N when selecting a letter at random from those in the word BANANA?

3. A card is drawn at random from a standard deck of 52 playing cards. Find the probability that the card is a spade.

4. In order to choose a mascot for a new school, the 1847 students were surveyed: 529 chose a falcon, 762 chose a ram, and 501 chose a panther. The remaining students did not vote. If a student is chosen at random, what is the probability that the student's choice was a panther?

© D.C. Heath and Company

Name _____

What you should learn :

15.2 How to find probabilities using the Fundamental Counting Principle or permutations.

Examples | *Using the Fundamental Counting Principle and Permutations*

a. An automobile license plate is made using two letters followed by three digits. How many license plates are possible?

This experiment has 5 events. The first event is the choice of a letter (out of 26). The second event is also the choice of a letter. The last three events are each the choice of a digit (out of 10). Using the Fundamental Counting Principle, you can conclude that there are

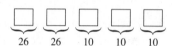

$26 \cdot 26 \cdot 10 \cdot 10 \cdot 10 = 676{,}000$

different license plates possible.

b. For next year's schedule of classes, mathematics, English, social studies, and science are each scheduled during the first four periods of the day. You have signed up for all four classes. If your schedule is randomly selected by a computer, what is the probability that your schedule will have English first period, math second period, science third period, and social studies fourth period?

Because there are four subjects, you have four choices for the first period. That leaves three choices for the second period, two choices for the third period, and only one choice for the fourth period. Because there is only one way to schedule your classes with English first, math second, science third, and social studies fourth, the probability of getting that schedule is

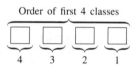

Order of first 4 classes

$$P = \frac{\text{One way to obtain desired schedule}}{\text{Number of ways to schedule four classes}} = \frac{1}{4 \cdot 3 \cdot 2 \cdot 1} = \frac{1}{24}.$$

Guidelines:
- To calculate the number of ways that several events can occur, use the Fundamental Counting Principle as stated on page 788 of the textbook.
- The number of permutations of *n* elements is *n*!.
- For permutation, order is important.

EXERCISES

1. Determine the number of possible 5-digit ZIP codes.

2. The members of a class of 9 students line up single file for lunch. In how many different orders can this occur?

3. Determine the possible number of 7-digit local telephone numbers that can be formed. (A local telephone number cannot have a 0 or 1 as its first or second digit.)

4. Seven letters are chosen, one at a time, at random from those in the word ENGLISH. Find the probability that they will be chosen in alphabetical order.

Name _____

What you should learn :

15.3	How to count and use combinations.

Correlation to Pupil's Textbook:

Mid-Chapter Test (p. 804)   Chapter Test (p. 828)
Exercises 8–12   Exercises 3, 4

Examples *Counting and Using Combinations*

a. A yogurt shop has a choice of ten toppings. In how many ways can you choose three different toppings for your yogurt?

Order is not important, so the number of ways is given by the number of combinations of 10 elements taken 3 at a time.

$$\binom{10}{3} = \frac{10!}{7!3!} = 120$$

There are 120 different combinations.

b. Of 458 seniors, 32 are taking calculus. If four students are randomly chosen from the senior class, what is the probability that all four are taking calculus?

To find the number of ways that four calculus students can be chosen, calculate the number of ways to choose 4 students from 32 students.

$$\binom{32}{4} = \frac{32!}{28!4!} = 35,960$$

To find the number of ways that four students can be chosen from the senior class, calculate the number of ways to choose 4 students from 458 students.

$$\binom{458}{4} = \frac{458!}{454!4!} = \frac{458 \cdot 457 \cdot 456 \cdot 455}{4!} = 1,809,450,370$$

Thus, the probability that all four students are taking calculus is

$$P = \frac{35,960}{1,809,450,370} \approx 0.00002.$$

Guidelines: To use combinations:
- The number of combinations of n elements taken m at a time is $\binom{n}{m}$.
- For combinations, order is not important.

EXERCISES

1. A box contains 8 different colored scarfs. Three scarfs are chosen from the box. How many color combinations are possible?

2. Out of a group of ten boys, three have blue eyes and five have brown eyes. If two of the boys are selected at random, what is the probability that both have blue eyes?

3. In how many ways can 4 girls be picked from a group of 30 girls?

4. A high school needs four additional faculty members: two math teachers, a chemistry teacher, and a French teacher. In how many ways can these positions be filled if there are six applicants for mathematics, two applicants for chemistry, and ten applicants for French?

Name _____

Correlation to Pupil's Textbook:

15.4	How to use unions, complements, and intersections to find the probability of an event.

Chapter Test (p. 828)
Exercises 6–8, 11, 12

Examples	*Using Unions, Complements, and Intersections*

a. A parking lot has 25 cars. Eight are red and 13 are have four doors. Six are both red and have four doors. Find the probability that a car chosen at random will be red or have four doors.

Let event A represent selecting a red car and let event B represent selecting a four-door car. Then, $P(A) = \frac{8}{25}$, $P(B) = \frac{13}{25}$, and $P(A \cap B) = \frac{6}{25}$. The probability that the car will be red *or* have four doors is

$$P(A \cup B) = P(A) + P(B) - P(A \cap B) \qquad \text{\textit{Probability of the union of two events}}$$

$$= \tfrac{8}{25} + \tfrac{13}{25} - \tfrac{6}{25} \qquad \text{\textit{Substitute for }} P(A),\ P(B),\ \text{\textit{and }} P(A \cap B).$$

$$= \tfrac{15}{25} = \tfrac{3}{5} \qquad \text{\textit{Simplify.}}$$

b. Students at Tech Memorial High School have three choices for a required science course in their junior year: physics, chemistry, or biology. Experience has shown that the probability of a student selecting physics is 0.12 and the probability of a student selecting chemistry is 0.57. If each student can select only one science course, what is the probability that a student will select biology?

Let event A represent selecting physics, event B represent selecting biology, and event C represent selecting chemistry.

$$P(B) = 1 - P(A \cup C) \qquad \text{\textit{B is the complement of }} A \cup C.$$

$$= 1 - [P(A) + P(C)] \qquad \text{\textit{A and C are mutually exclusive events.}}$$

$$= 1 - [0.12 + 0.57] \qquad \text{\textit{Substitute for }} P(A) \text{ \textit{and }} P(C).$$

$$= 1 - 0.69 = 0.31 \qquad \text{\textit{Simplify.}}$$

Guidelines:
- The probability of $A \cup B$ is $P(A \cup B) = P(A) + P(B) - P(A \cap B)$.
- If A and B are mutually exclusive, then $P(A \cup B) = P(A) + P(B)$.
- $P(A') = 1 - P(A)$, where A' is the complement of A.

EXERCISES

1. If $P(A) = \frac{6}{11}$, $P(B) = \frac{8}{11}$, and $P(A \cap B) = \frac{5}{11}$, find $P(A \cup B)$.

2. If $P(A) = \frac{13}{20}$, $P(B) = \frac{7}{20}$, and $P(A \cup B) = \frac{9}{10}$, find $P(A \cap B)$.

3. If $P(A) = \frac{6}{11}$, find $P(A')$.

4. If one card is chosen at random from a standard deck of 52 playing cards, what is the probability that the card is an ace or a club?

5. A card is drawn at random from a standard deck of 52 playing cards. What is the probability that the card is *not* an ace?

Reteach
Chapter 15

Name _____

What you should learn :

15.5	How to find the probability of independent events.

Correlation to Pupil's Textbook:

Chapter Test (p. 828)
Exercise 9

Examples | *Finding the Probability of Independent Events*

a. Two integers from 0 through 9 are chosen by a random number generator. What is the probability of choosing the number 2 both times?

Let event A represent selecting the first number 2, and let event B represent selecting the second number 2. Then $P(A) = \frac{1}{10}$ and $P(B) = \frac{1}{10}$. The probability of selecting the number 2 both times is

$$P(A \cap B) = P(A) \cdot P(B) \qquad \text{\textit{A and B are independent events.}}$$
$$= \frac{1}{10} \cdot \frac{1}{10} \qquad \text{\textit{Substitute } } \frac{1}{10} \text{ \textit{for} } P(A) \text{ \textit{and} } \frac{1}{10} \text{ \textit{for} } P(B).$$
$$= \frac{1}{100} \qquad \text{\textit{Multiply.}}$$

b. The probability of selecting a rotten apple from a basket of apples is 12%. What is the probability of selecting three good apples when selecting one from each of three different baskets?

Let event A represent selecting a rotten apple. Then A' represents selecting a good apple. Because $P(A) = 0.12$, it follows that $P(A') = 0.88$. The probability of selecting three good apples is as follows.

$$P(A' \cap A' \cap A') = P(A') \cdot P(A') \cdot P(A') \qquad \text{\textit{The events are independent.}}$$
$$= (0.88)(0.88)(0.88) \qquad \text{\textit{Substitute 0.88 for } } P(A').$$
$$= (0.88)^3 \qquad \text{\textit{Simplify.}}$$
$$\approx 0.681 \qquad \text{\textit{Use a calculator.}}$$

Guidelines: • If A and B are independent events, then the probability that both A and B occur is $P(A \cap B) = P(A) \cdot P(B)$.

EXERCISES

1. A die is tossed three times. What is the probability that a 2 will come up all three times?

2. Two cards are randomly selected from a standard deck of 52 playing cards. If the first card is replaced before the second is drawn, what is the probability that the first card will be a jack and the second card will be a queen?

3. Each of 20 questions on a multiple choice test has 5 possible choices. If a student randomly selected his answers for each of the 20 questions, what is the probability that he will answer every question correctly?

4. In Exercise 2, assume that the first card is not replaced before the second is drawn. Now, what is the probability that the first card will be a jack and the second card will be a queen?

Reteach

Chapter 15

Name _____

What you should learn :

| 15.6 | How to find the expected value of a sample space. |

Correlation to Pupil's Textbook:

Chapter Test (p. 828)
Exercise 10

Examples | *Finding the Expected Value of a Sample Space*

a. A sample space has five outcomes, each with the indicated payoff and probability. Find the expected value.

Payoff	$1	$2	$3	$4	$5
Probability	0.15	0.23	0.21	0.30	0.11

$$V = P_1x_1 + P_2x_2 + P_3x_3 + P_4x_4 + P_5x_5 \qquad \textit{Formula for expected value}$$
$$= (0.15)(1) + (0.23)(2) + (0.21)(3) + (0.30)(4) + (0.11)(5) \qquad \textit{Substitute.}$$
$$= 2.99 \qquad \textit{The expected value is \$2.99.}$$

b. An integer is chosen at random from 1 through 9. If the number 7 is selected, the player wins 5 points. If the number 3 is selected, the player wins 2 points. Otherwise, the player loses 1 point. What is the expected point value? Is the game fair?

Let A represent selecting a 7. Let B represent selecting a 3, and let C represent selecting a number that is not a 7 or a 3. Then, $P(A) = \frac{1}{9}$, $P(B) = \frac{1}{9}$, and $P(C) = \frac{7}{9}$. The expected value is

$$V = P(A) \cdot (5) + P(B) \cdot (2) + P(C) \cdot (-1) \qquad \textit{Formula for expected value}$$
$$= (\tfrac{1}{9})(5) + (\tfrac{1}{9})(2) + (\tfrac{7}{9})(-1) \qquad \textit{Substitute.}$$
$$= \tfrac{5}{9} + \tfrac{2}{9} - \tfrac{7}{9} = 0. \qquad \textit{The game is fair because } V = 0.$$

Guidelines: To find the expected value of a sample space:
- Use the formula on page 817 of the textbook.
- If the expected value is 0, the game is fair.

EXERCISES

1. A sample space has five outcomes, each with the indicated payoff and probability. Find the expected value of the sample space.

Payoff	$1	$2	$3	$4	$5
Probability	0.13	0.29	0.11	0.24	0.23

2. An integer is chosen at random from 1 through 9. If the number 7 is selected, the player wins 5 points. Otherwise, the player loses 1 point. Find the expected value. Is the game fair?

3. An integer is chosen at random from the digits 1 through 9. If the number is a 3 or a 5, the player wins 5 points. Otherwise the player loses 2 points. Find the expected value. Is the game fair?

© D. C. Heath and Company

Answers to Exercises

Lesson 1.1

1. $\frac{1}{3}, 1, \sqrt{2}$ **2.** $-1, \frac{3}{5}, 1$ **3.** $\frac{2}{3}, \sqrt{5}, 3.25$

4. $-4, -\frac{5}{2}, 1$ **5.** $\frac{3}{4}, \frac{4}{3}, 2$ **6.** $-\sqrt{3}, -\frac{1}{2}, 0$

7. $-2, -\frac{2}{5}, 0, \frac{1}{3}, 1$ **8.** $-15, -\sqrt{2}, \frac{1}{2}, 3, 5.7$

9. $-1, -\frac{1}{2}, 0, 2, \sqrt{7}$ **10.** $-5, -\frac{5}{4}, 1, \frac{5}{2}, \frac{10}{3}$

11. $-\sqrt{3}, -0.3, \frac{1}{2}, \sqrt{2}, 4$ **12.** $-5, -\sqrt{5}, 0, 3, 4$

Lesson 1.2

1. 17 **2.** 2 **3.** 9 **4.** 60 **5.** 35

6. 2 **7.** 61 **8.** 28 **9.** $\frac{17}{4}$ **10.** 248

Lesson 1.3

1. -6 **2.** -24 **3.** 4 **4.** $\frac{4}{3}$ **5.** 12

6. 1 **7.** 3 **8.** -19 **9.** 4 **10.** $\frac{2}{5}$

11. 0 **12.** $-\frac{5}{4}$ **13.** $-\frac{1}{2}$ **14.** $\frac{1}{9}$ **15.** 1

Lesson 1.4

1. 28 **2.** 20 minutes

Lesson 1.5

1. $r = \dfrac{A}{2\pi h}$ **2.** $w = \dfrac{V}{lh}$ **3.** $b = \frac{3}{4}h$

4. $h = \dfrac{3V}{\pi r^2}$ **5.** $r = \dfrac{L - S}{L}$

6. $t = \dfrac{A - P}{Pr}$ **7.** $x = 1000(12 - p)$

8. $h = \dfrac{3V}{b^2}$ **9.** $x = \dfrac{C - 5000}{0.56}, 6000$

Lesson 1.6

1. $x < \frac{1}{2}$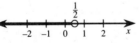

2. $x \geq -2$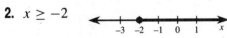

3. $x < -3$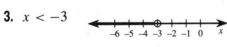

4. $x \geq \frac{1}{2}$

5. $x > 0$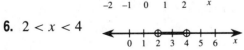

6. $2 < x < 4$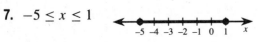

7. $-5 \leq x \leq 1$

8. $x \leq -1$ or $x \geq 1$

9. $-6 < x < 4$

Lesson 1.7

1. $\frac{6}{5}, 2$ **2.** $-1, -\frac{9}{17}$ **3.** $0 < x < \frac{5}{2}$

4. $-2 \leq x \leq 1$ **5.** $y < -8$ or $y > 2$

6. $x \leq -10$ or $x \geq 18$

Lesson 1.8

1.

Number	Tally	Frequency
0	|||	3
1	⑊⑊ ||	7
2	⑊⑊ ⑊⑊	10
3	⑊⑊ |	6
4	|	1
5	|	1

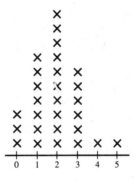

2.

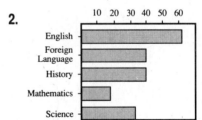

Lesson 2.1

1.

2.

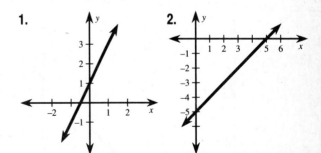

3.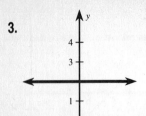

4.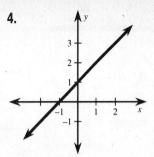

15.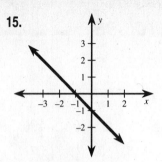

5.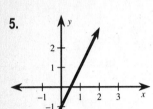

6.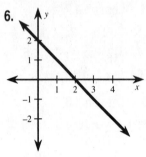

1. $\frac{3}{2}$ **2.** 1   **3.** $-\frac{1}{2}$ **4.** Undefined
5. 1   **6.** −1   **7.** 0   **8.** 2
9. Perpendicular   **10.** Neither

■ **Lesson 2.3**

1.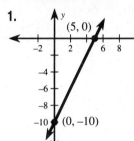

2.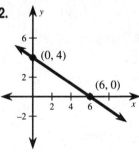

7.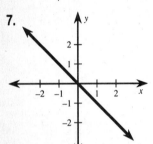

8.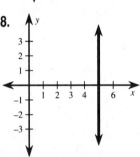

3.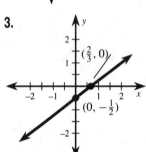

4.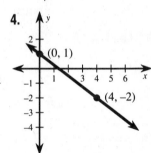

9.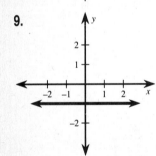

10.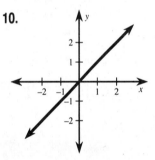

5.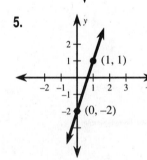

6.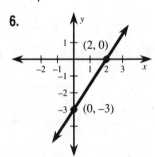

11.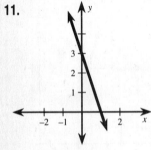

12.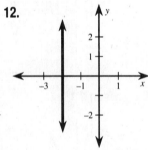

■ **Lesson 2.4**

1. $y = 4x + \frac{2}{3}$ **2.** $y = \frac{1}{2}x + \frac{1}{2}$
3. $y = -x + 1$ **4.** $y = \frac{2}{3}x - \frac{7}{3}$
5. $y = -\frac{1}{3}x - \frac{2}{3}$ **6.** $y = -\frac{1}{4}x - 2$
7. $y = \frac{1}{3}x + 1$ **8.** $y = -2x - 5$
9. $y = -\frac{3}{2}x - \frac{3}{2}$

13.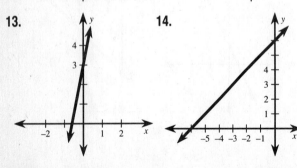

14.

Lesson 2.5

1.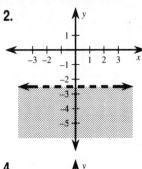

2.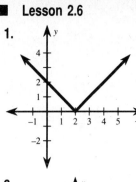

3.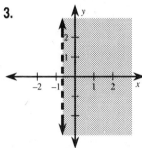

4.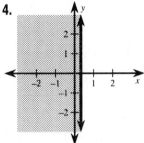

5.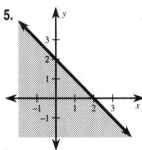

6.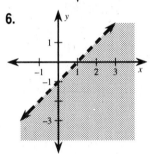

7.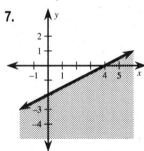

8.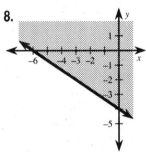

9.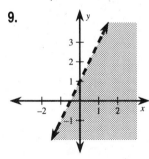

10.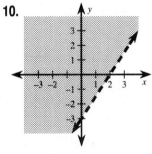

11.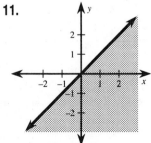

12.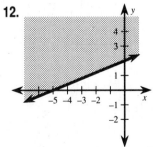

Lesson 2.6

1.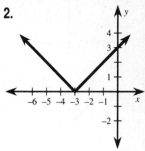

2.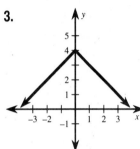

3.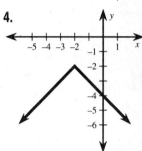

4.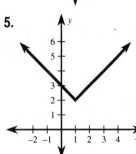

5.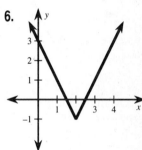

6.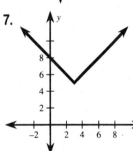

7.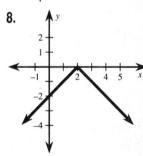

8.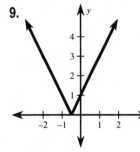

9.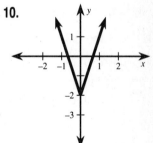

10.

11.

12.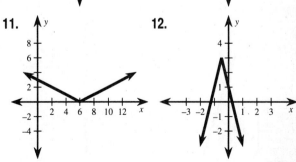

■ **Lesson 2.7**

1. $y = -2x + \frac{15}{2}$ 2. $y = \frac{8}{3}x + 24$

■ **Lession 3.1**

1. $(4, 0)$ 2. $(3, -3)$

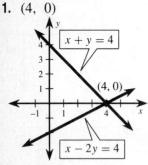

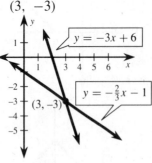

3. $\left(-\frac{1}{2}, 3\right)$ 4. No solution

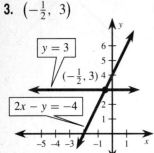

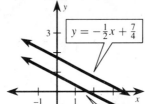

5. Many solutions   6. $(2, 4)$

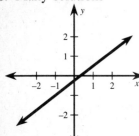

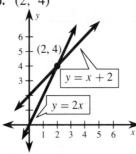

■ **Lesson 3.2**

1. $\left(\frac{1}{2}, -5\right)$ 2. $(2, 3)$ 3. $(-1, 3)$
4. $\left(\frac{1}{2}, 1\right)$ 5. $(5, -9)$ 6. No solution

■ **Lesson 3.3**

1. 26 ft by 18 ft

2. 370 letters were mailed at $0.29 each.
   298 letters were mailed at $0.52 each.

■ **Lesson 3.4**

1. $(1, 1)$, $(1, 2)$, $\left(\frac{4}{3}, 1\right)$ 2. $(0, -3)$, $(4, 1)$

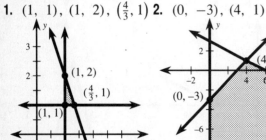

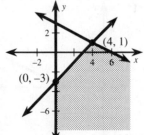

3. $(0, 2)$, $(2, 2)$,
   $(2, 0)$

4. $(1, 2)$, $(1, 6)$,
   $(4, 2)$, $(4, 6)$

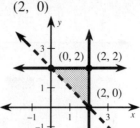

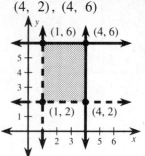

■ **Lesson 3.5**

1. 0, 17   2. 7, 28   3. 7, 13

■ **Lesson 3.6**

1. $(2, -2, 3)$ 2. $(-1, 3, 1)$ 3. $(1, -1, 3)$

■ **Lesson 4.1**

1. $\begin{bmatrix} -1 & -2 \\ -3 & 0 \end{bmatrix}$ 2. $\begin{bmatrix} 1 & 6 \\ -4 & 3 \\ 5 & -6 \end{bmatrix}$

3. $\begin{bmatrix} -3 & 0 & 3 \\ -9 & -1 & 9 \\ 3 & 1 & 0 \end{bmatrix}$ 4. $\begin{bmatrix} 4 & 20 & 30 \\ -2 & 6 & -12 \end{bmatrix}$

5. $\begin{bmatrix} -6 & -1 \\ 5 & 3 \end{bmatrix}$ 6. $\begin{bmatrix} 7 & 18 & -13 \\ 2 & 11 & -8 \end{bmatrix}$

■ **Lesson 4.2**

1. $\begin{bmatrix} 3 & 4 & 1 \\ 17 & 8 & 3 \\ 7 & -9 & -1 \end{bmatrix}$ 2. $\begin{bmatrix} 4 & 12 \\ -15 & 11 \end{bmatrix}$

3. $\begin{bmatrix} 1 & 1 & 1 \\ 7 & 7 & 6 \end{bmatrix}$ 4. $[\,18 \quad 19\,]$

5. $[-2]$ 6. $\begin{bmatrix} -2 & 15 & -5 \\ 27 & -32 & 3 \\ 0 & -7 & 2 \end{bmatrix}$

■ **Lesson 4.3**

1. -8 2. 14   3. 0   4. 29
5. -7 6. -24

■ **Lesson 4.4**

1. $\begin{bmatrix} -5 & -7 \\ 2 & 3 \end{bmatrix}$ 2. $\begin{bmatrix} \frac{11}{2} & -\frac{5}{2} \\ -2 & 1 \end{bmatrix}$

3. $\begin{bmatrix} \frac{2}{3} & -\frac{1}{3} \\ -\frac{1}{3} & \frac{2}{3} \end{bmatrix}$ 4. $\begin{bmatrix} 1 & 4 \\ 5 & -3 \end{bmatrix}$

5. $\begin{bmatrix} 3 & 3 & 1 \\ 1 & 2 & 1 \end{bmatrix}$

■ **Lesson 4.5**

1. $(2, 3)$ 2. $(-8, 5)$ 3. $(3, 3)$
4. $(32, 3, -8)$

■ Lesson 4.6

1. $(1, -1, 2)$ **2.** $\left(-\frac{1}{2}, -\frac{1}{2}, -2\right)$

■ Lesson 4.7

1. $(-1, 3)$ **2.** $(2, -1)$ **3.** $(6, -2)$
4. $(4, 0, -3)$ **5.** $(2, 1, -1)$ **6.** $(5, -1, 2)$

■ Lesson 5.1

1. ± 8 **2.** $\pm\sqrt{13}$ **3.** $\pm\sqrt{32}$ **4.** ± 5
5. $\pm\sqrt{45}$ **6.** ± 5 **7.** ± 12 **8.** ± 6
9. ± 6 **10.** 9.22 in. **11.** 5.86 sec

■ Lesson 5.2

1. 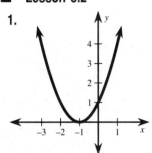 **2.**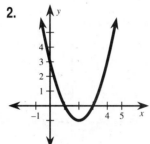

3. 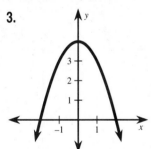 **4.**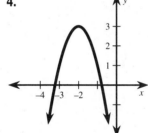

5. 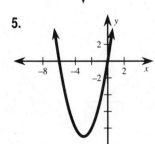 **6.**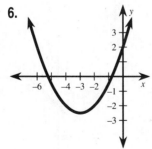

■ Lesson 5.3

1. $-7, 1$ **2.** $5 - \sqrt{14}, 5 + \sqrt{14}$
3. $\frac{1}{2} - \frac{\sqrt{5}}{2}, \frac{1}{2} + \frac{\sqrt{5}}{2}$ **4.** $-\frac{9}{2}, -\frac{1}{2}$
5. $2 - \frac{\sqrt{3}}{2}, 2 + \frac{\sqrt{3}}{2}$ **6.** $\frac{1}{4} - \frac{\sqrt{17}}{4}, \frac{1}{4} + \frac{\sqrt{17}}{4}$

■ Lesson 5.4

1. $\frac{-5-\sqrt{17}}{2}, \frac{-5+\sqrt{17}}{2}$ **2.** $-\frac{1}{2}, 2$
3. $\frac{5-\sqrt{53}}{14}, \frac{5+\sqrt{53}}{14}$ **4.** $\frac{1}{2} - \frac{\sqrt{21}}{6}, \frac{1}{2} + \frac{\sqrt{21}}{6}$
5. $\frac{-7-\sqrt{97}}{8}, \frac{-7+\sqrt{97}}{8}$ **6.** $\frac{1-\sqrt{13}}{2}, \frac{1+\sqrt{13}}{2}$
7. $\frac{3}{2}$ **8.** $-2 - \frac{\sqrt{8}}{2}, -2 + \frac{\sqrt{8}}{2}$ **9.** $-\frac{15}{2}, \frac{9}{2}$

■ Lesson 5.5

1. $-4i$ **2.** $-i$ **3.** $\sqrt{2}$ **4.** $2 + 2i$
5. $-5 - i$ **6.** $9 - 2i$ **7.** -3 **8.** 85
9. $11 + 7i$ **10.** $-2 + 24i$ **11.** $64 + 41i$
12. $-21 + 20i$

■ Lesson 5.6

1. $-\frac{5}{2} - \frac{\sqrt{3}}{2}i, -\frac{5}{2} + \frac{\sqrt{3}}{2}i$
2. $-1 - \frac{\sqrt{12}}{2}i, -1 + \frac{\sqrt{12}}{2}i$ **3.** $2 - 3i, 2 + 3i$
4. $1 + \frac{i\sqrt{8}}{2}, 1 - \frac{i\sqrt{8}}{2}$
5. $\frac{1}{2} - \frac{\sqrt{20}}{4}i, \frac{1}{2} + \frac{\sqrt{20}}{4}i$
6. $-\frac{1}{2} - \frac{\sqrt{3}}{2}i, -\frac{1}{2} + \frac{\sqrt{3}}{2}i$
7. $-\frac{1}{6} - \frac{\sqrt{23}}{6}i, -\frac{1}{6} + \frac{\sqrt{23}}{6}i$
8. $\frac{3}{2} - \frac{\sqrt{11}}{2}i, \frac{3}{2} + \frac{\sqrt{11}}{2}i$
9. $\frac{3}{14} - \frac{\sqrt{411}}{14}i, \frac{3}{14} + \frac{\sqrt{411}}{14}i$

■ Lesson 5.7

1. 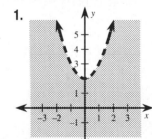 **2.**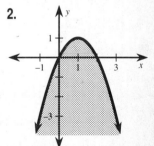

3. **4.**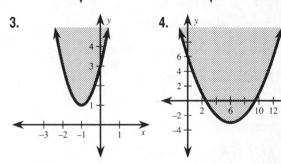

5.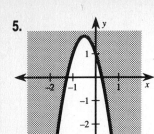

6.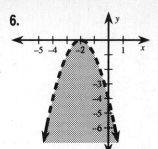

7.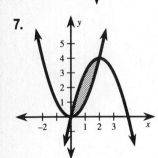

8.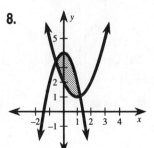

9.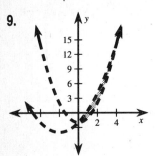

■ Lesson 6.1

1. It is not.　　**2.** It is.　　**3.** It is not.

4. It is.　　**5.** 18　　**6.** 0　　**7.** 3　　**8.** −2

■ Lesson 6.2

1. $2 - 4x$, all real numbers

2. $6x - 3x^2$, all real numbers

3. $\dfrac{2 - x}{3x}$, all real numbers except 0

4. $12x$, all real numbers

5. $2x + 2$, all real numbers

6. $\dfrac{3x}{2 - x}$, all real numbers except 2

7. $2 - 3x$, all real numbers

8. $6 - 3x$, all real numbers

■ Lesson 6.3

1. $g(x) = x + 3$　　**2.** $g(x) = \frac{1}{3}x + \frac{2}{3}$

3. $g(x) = 5 - x$　　**4.** $g(x) = \frac{1}{4}x - \frac{3}{4}$

5. $g(x) = \frac{3}{2} - \frac{1}{2}x$　　**6.** $g(x) = 2x + 4$

■ Lesson 6.4

1. −7　　**2.** −5　　**3.** −3　　**4.** −24

5.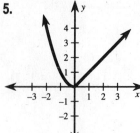

6.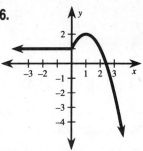

7. $f(x) = \begin{cases} -x + 3, & x < 3 \\ x - 3, & x \geq 3 \end{cases}$

8. $f(x) = \begin{cases} -\frac{1}{2}x - 1, & x < -2 \\ \frac{1}{2}x + 1, & x \geq -2 \end{cases}$

■ Lesson 6.5

1.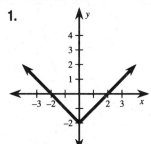

2.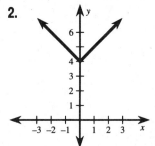

3.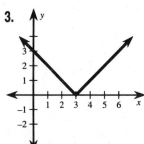

4.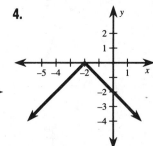

5.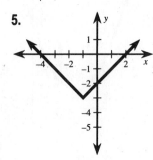

6.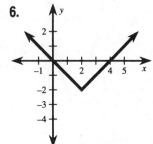

7.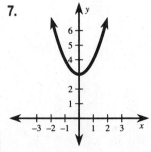

8.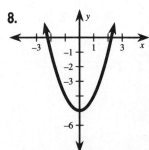

9.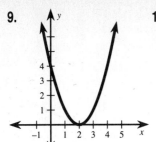

10.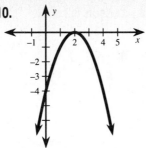

11.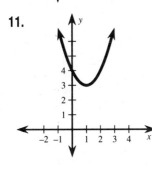

12.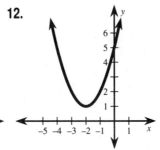

■ Lesson 6.6

1. $10, 8, 6, 4, 2, 0, -2; \ f(n) = -2n + 12$

2. $1, 3, 7, 13, 21, 31, 43; \ f(n) = n^2 + n + 1$

3. $2, 3, 5, 8, 12, 17, 23; \ f(n) = \frac{1}{2}n^2 - \frac{1}{2}n + 2$

4. $3, 5, 7, 9, 11, 13, 15; \ f(n) = 2n + 1$

■ Lesson 6.7

1. Mean $= 76$
Median $= 78.5$
Mode $= 75$

2. $59.5, 78.5, 91$

3.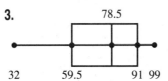

■ Lesson 7.1

1. 8   **2.** -5832 **3.** t^5 **4.** $\frac{1}{16}$

5. $\dfrac{1}{9x^4y^2}$ **6.** $\frac{2}{9}$ **7.** $\dfrac{y^5}{x^4}$ **8.** $-\frac{3}{2}x^8y^5$

9. $-\dfrac{2}{xy^5}$ **10.** -3 **11.** 5 **12.** $\pm\frac{1}{4}$

■ Lesson 7.2

1. $\$597.81$ **2.** $\$2835.25$ **3.** $\$764.15$

4.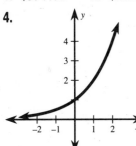

5.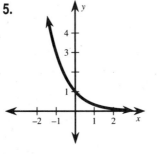

■ Lesson 7.3

1. 9   **2.** 7   **3.** -8 **4.** $\frac{1}{27}$ **5.** 64

6. $\frac{1}{216}$ **7.** $7^{1/2}$ **8.** $(-17)^{1/3}$ **9.** $180^{1/9}$

■ Lesson 7.4

1. $2^{3/2}$ **2.** $6^{3/5}$ **3.** $3^{1/2}$ **4.** $2x^2$

5. 6   **6.** $\frac{1}{4}$ **7.** $3\sqrt[3]{2}$ **8.** 2

9. 7   **10.** $3^{1/6}$ **11.** $-2y$ **12.** $6x^8$

13. $4x^2\sqrt[4]{y}$ **14.** $4|x|$ **15.** $5^{9/4}$

■ Lesson 7.5

1. 16   **2.** 16   **3.** ±11 **4.** 32

5. 24   **6.** 1   **7.** 4   **8.** 5

9. -1 **10.** 7   **11.** $\frac{1}{5}$ **12.** No solution

■ Lesson 7.6

1. Domain: $x \geq -3$
Range: $y \geq 0$

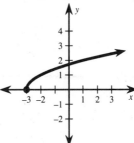

2. Domain: $x \geq 0$
Range: $y \geq 3$

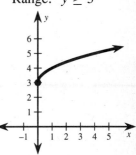

3. Domain: $x \geq -3$
Range: $y \leq 0$

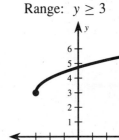

4. Domain: $x \geq -3$
Range: $y \geq 3$

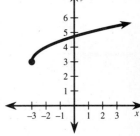

5. Domain:
Real numbers
Range:
Real numbers

6. Domain:
Real numbers
Range:
Real numbers

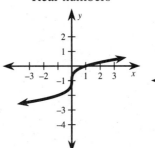

7. Domain:
   Real numbers
   Range:
   Real numbers

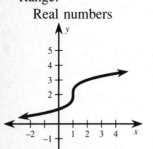

8. Domain:
   Real numbers
   Range:
   Real numbers

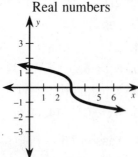

9. Domain: $x \geq 0$
   Range: $y \leq 2$

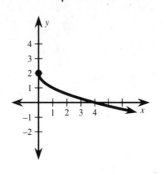

■ **Lesson 8.1**

1.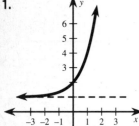

2.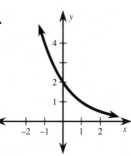

3.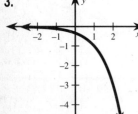

4.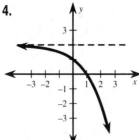

5.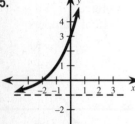

6.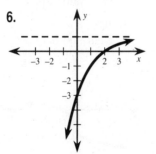

7. $331.95

■ **Lesson 8.2**

1. 1   **2.** 2   **3.** 0   **4.** −4   **5.** $\frac{1}{2}$

6. 4   **7.** 6   **8.** 3   **9.** 4

10.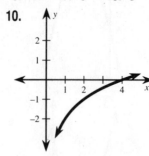

11.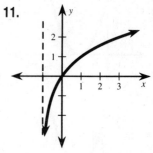

12.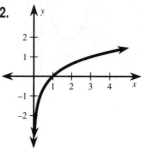

■ **Lesson 8.3**

1. $\log_2 3 + 2\log_2 x$ **2.** $1 + \log_4 y$ **3.** -1

4. $\log_{10} x + \log_{10} y + \log_{10} z$

5. $2\log_5 x - \log_5 y$

6. $\frac{1}{2}\log_2 x - \log_2 y - \log_2 z$

7. $\log_5 3x^2$ **8.** $\log_{10} \dfrac{y}{8}$ **9.** $\log_4 \dfrac{x^5 z^3}{y^2}$

■ **Lesson 8.4**

1. e^{11} **2.** e^3 **3.** $\dfrac{2}{e^2}$

4. $27e^6$ **5.** $4e^{18}$ **6.** e^{2x+2}

7.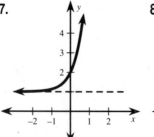

8.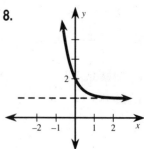

9. $598.61   **10.** $637.63

■ **Lesson 8.5**

1. $\frac{1}{4}\ln x$ **2.** $\ln 2 + 3\ln x + 6\ln y$

3. $\frac{1}{3}[2\ln a + \ln b - \ln c]$

4. $\ln 3 + 2\ln y - \ln x$ **5.** $\ln x - \ln y - \ln z$

6. $2\ln 2 + 6\ln x$ **7.** $\ln 3x^4$ **8.** $\ln \dfrac{x\sqrt{y}}{z}$

9. $\ln \dfrac{x^2 y^2}{z}$ **10.** 17 **11.** -1 **12.** $\ln 6 + 2$

13.

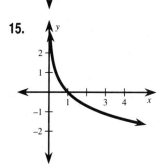

14.

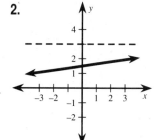

15.

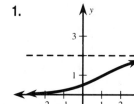

■ **Lesson 8.6**

1. $\log_3 8.4 \approx 1.937$ **2.** $\tfrac{1}{2}\ln 17 \approx 1.417$
3. $\log_2 4 = 2$ **4.** $-\ln 15 \approx -2.708$
5. $\tfrac{1}{3}\ln 18 \approx 0.963$ **6.** $\log_{10} 16.6 \approx 1.220$
7. 0 **8.** $e^{1/3} \approx 1.396$ **9.** $\tfrac{1}{2}e^{5/4} \approx 1.745$
10. 81 **11.** $10^{-2/13} \approx 0.702$
12. $e^3 \approx 20.086$

■ **Lesson 8.7**

1.

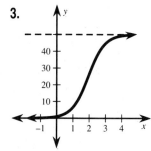

2.

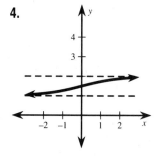

3.

4.

5.

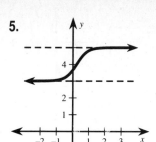

6.

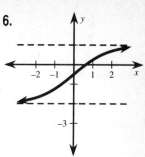

7. 294.60 **8.** 4.76

■ **Lesson 9.1**

1. $7x^3 - 15x^2 - 2x + 8$ **2.** $x^2 - 4x - 2$
3. $3x^3 + 3x^2 - 6x$ **4.** $2x^3 + 2x^2 - 2x$
5. $2x^2 - x - 10$ **6.** $4x^2 - 12x + 9$
7. $x^3 + x^2 - 7x - 3$ **8.** $x^3 - 8$
9. $x^2 - 25$ **10.** $8x^3 - 12x^2 + 6x - 1$

■ **Lesson 9.2**

1.

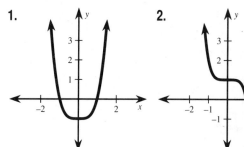

2.

3.

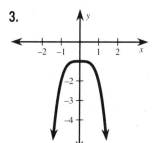

4.

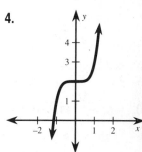

5.

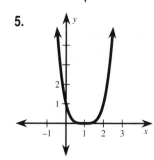

6.

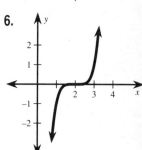

7.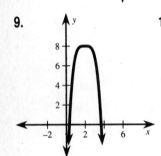

8.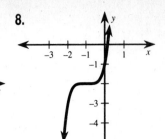

9.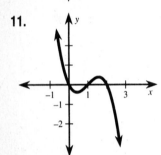

10.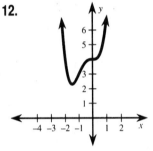

11.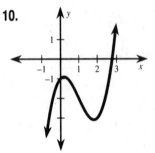

12.

Lesson 9.5

1. $-3, 1, 2$ **2.** $-4, 1, 2$ **3.** $-1, -\frac{1}{2}, 3$

4. $-3, \pm\sqrt{5}$ **5.** $1, 2 \pm \sqrt{3}$

6. $-1, 1 \pm \sqrt{2}$ **7.** $-3, \dfrac{-5 \pm \sqrt{17}}{2}$

8. $-2, -\frac{3}{2}, \pm 1$

Lesson 9.6

1. $(x - 1)(x - 5)(x - 3i)(x + 3i)$

2. $(x - 2)(x - 3)(x - i)(x + i)$

3. $(x + 1)(x - \sqrt{6})(x + \sqrt{6})$

4. $(x - 4)(x - 2i)(x + 2i)$

5. $(x - 1)(x + 4)(x - \sqrt{2}i)(x + \sqrt{2}i)$

6. $(x + 2)(x - 2)(x + 2i)(x - 2i)$

Lesson 9.7

1. $16, \approx 4.8$ **2.** $2.5, \approx 0.69$

Lesson 10.1

1. $y = 1, x = 3$ **2.** $y = 0, x = -1$

3. $y = 3$ **4.** $x = 2$

5.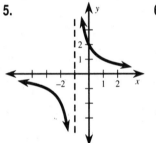

6.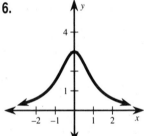

7.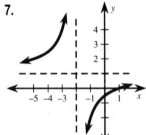

8.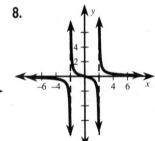

Lesson 9.3

1. $27x(x - 3)$ **2.** $4x^2(2x - 3)$

3. $(x - 2)(x^2 - 2)$

4. $(4x - 3)(16x^2 + 12x + 9)$

5. $5x(2x + 3)^2$ **6.** $(2x + 3)(4x^2 - 6x + 9)$

7. $0, 3$ **8.** $-3, 5$ **9.** $0, -\frac{2}{3}, \frac{2}{3}$

10. $-6, 4$ **11.** $-6, -1, 1$ **12.** $\frac{1}{2}$

Lesson 9.4

1. $x^2 - x - 1 + \dfrac{5}{2x + 1}$ **2.** $x^2 - 4$

3. $x^2 - 8x + 13 - \dfrac{25}{x + 2}$

4. $5x^3 - 5x^2 + 3x - 3 + \dfrac{4}{x + 1}$

5. $(x - 4)(x + 1)(3x - 2)$

6. $(x + 2)(x + 1)(x - 4)$

Lesson 10.2

1. $y = \dfrac{5}{x}, \dfrac{5}{2}$ **2.** $y = -\dfrac{9}{x}, -\dfrac{9}{2}$ **3.** $y = \dfrac{4}{x}, 2$

4. $z = 2xy$ **5.** $z = -\frac{1}{3}xy$ **6.** $z = 32xy$

Lesson 10.3

1. $\dfrac{8x}{5}$ 2. $\dfrac{3}{4x^2}$ 3. $\dfrac{x-2}{x+1}$

4. $\dfrac{x(x+3)}{x+1}$ 5. $\dfrac{t^3}{2}$ 6. $\dfrac{20x}{3y^3}$

7. $3x^2$ 8. $\dfrac{x}{3(x-1)}$ 9. $\dfrac{x+6}{4(x+2)}$

Lesson 10.4

1. $-\dfrac{1}{3}$ 2. No solution 3. $-\dfrac{12}{7}$ 4. -4

5. No solution 6. $\dfrac{9}{11}$ 7. -4 8. 8

9. 3 10. $-3, \dfrac{4}{3}$ 11. -2 12. $-1, 3$

Lesson 10.5

1. $\dfrac{15x+2}{3x^2}$ 2. $\dfrac{x^2+4x-6}{2x^2}$

3. $\dfrac{-x-17}{(x+5)(x+1)}$ 4. $\dfrac{x+6}{x^2-4}$

5. $\dfrac{1}{x+1}$ 6. $\dfrac{x^2+4x}{x^2-1}$

7. $\dfrac{x}{3(x-1)}$ 8. $\dfrac{2x-1}{x^2}$ 9. $\dfrac{x+1}{x-1}$

Lesson 10.6

1. $332.14, $11,957.04, $1957.04

2. $643.70, $231,732, $151,732

3. $90.97, $1091.64, $91.64

Lesson 11.1

1. $x^2 = y$ 2. $x^2 = 4y$ 3. $y^2 = -2x$
4. $x^2 = -12y$ 5. $y^2 = -8x$
6. $y^2 = 2x$ 7. $x^2 = -4y$ 8. $y^2 = 16x$
9. $(0, \frac{1}{2})$, $y = -\frac{1}{2}$ 10. $(4, 0)$, $x = -4$

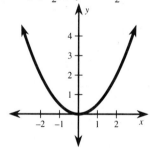
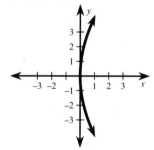

11. $(0, -\frac{1}{16})$, $y = \frac{1}{16}$ 12. $(-1, 0)$, $x = 1$

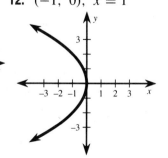

Lesson 11.2

1. $x^2 + y^2 = 25$ 2. $x^2 + y^2 = 82$
3. $x^2 + y^2 = 13$ 4. $x^2 + y^2 = 17$
5. $\left(\sqrt{2}, \sqrt{2}\right), \left(-\sqrt{2}, -\sqrt{2}\right)$
6. $\left(\frac{12}{5}, \frac{16}{5}\right), (-4, 0)$
7. $(2, 2), (-4, 8)$ 8. $\left(1, \frac{1}{2}\right), \left(9, -\frac{3}{2}\right)$

Lesson 11.3

1. $\dfrac{x^2}{9} + \dfrac{y^2}{1} = 1$ 2. $\dfrac{x^2}{9} + \dfrac{y^2}{16} = 1$

3. $\dfrac{x^2}{16} + \dfrac{y^2}{4} = 1$ 4. $\dfrac{x^2}{4} + \dfrac{y^2}{25} = 1$

5. Eccentricity $= \dfrac{\sqrt{5}}{3}$ 6. Eccentricity $= \dfrac{1}{\sqrt{2}}$

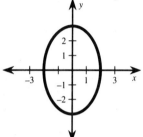
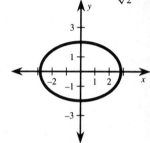

7. Eccentricity $= \dfrac{\sqrt{8}}{3}$ 8. Eccentricity $= \dfrac{5}{\sqrt{34}}$

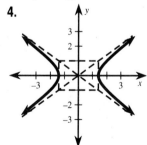
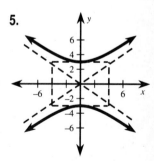

Lesson 11.4

1. $\dfrac{x^2}{4} - \dfrac{y^2}{21} = 1$ 2. $\dfrac{x^2}{4} - \dfrac{y^2}{5} = 1$

3. $\dfrac{y^2}{1} - \dfrac{x^2}{48} = 1$

4. 5.

6.

7.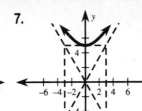

■ Lesson 11.5

1. $(y-1)^2 = 4(x+2)$

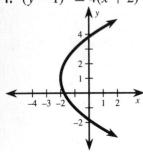

2. $\dfrac{(y+2)^2}{9} - \dfrac{(x-3)^2}{16} = 1$

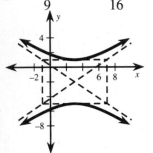

3. $(x+1)^2 + (y+3)^2 = 4$

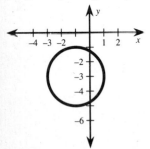

4. $(x-1)^2 + \dfrac{(y-2)^2}{4} = 1$

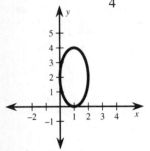

■ Lesson 11.6

1. Hyperbola   **2.** Parabola   **3.** Parabola
4. Ellipse   **5.** Circle   **6.** $(-1, -1)(3, 7)$
7. $\left(-\dfrac{\sqrt{3}}{2}, \dfrac{3}{2}\right), \left(\dfrac{\sqrt{3}}{2}, \dfrac{3}{2}\right)$ **8.** $(0, 2)$

■ Lesson 12.1

1. 2, 7, 12, 17, 22   **2.** $-1, 0, 1, 2, 3$
3. 6, 12, 24, 48, 96   **4.** $-1, \dfrac{1}{2}, -\dfrac{1}{3}, \dfrac{1}{4}, -\dfrac{1}{5}$
5. $\dfrac{1}{6}, \dfrac{1}{12}, \dfrac{1}{20}, \dfrac{1}{30}, \dfrac{1}{42}$ **6.** $\dfrac{1}{3}, \dfrac{1}{9}, \dfrac{1}{27}, \dfrac{1}{81}, \dfrac{1}{243}$
7. $\dfrac{1}{2}, \dfrac{2}{5}, \dfrac{3}{10}, \dfrac{4}{17}, \dfrac{5}{26}$ **8.** 2, 4, 2, 4, 2
9. 34   **10.** $\dfrac{9}{4}$ **11.** 8   **12.** 70

■ Lesson 12.2

1. $a_n = 9 - 4n$ **2.** $a_n = 21 - 2n$
3. $a_n = 3n + 6$
4. $a_n = \dfrac{3}{2}n + 12$ **5.** 84   **6.** 61   **7.** 114
8. 209   **9.** 13,500   **10.** 3,825

■ Lesson 12.3

1. $a_n = 2\left(-\dfrac{1}{3}\right)^{n-1}$ **2.** $a_n = 5(1.1)^{n-1}$
3. $a_n = 3(-1)^{n-1}$
4. $a_n = 8\left(\dfrac{1}{2}\right)^{n-1}$ **5.** $\dfrac{1}{16}$ **6.** ≈ 30.580
7. -64 **8.** ≈ 7.992 **9.** ≈ 2.883
10. ≈ 4.275

■ Lesson 12.4

1. 6   **2.** It has none.   **3.** $\dfrac{4}{3}$ **4.** $\dfrac{50}{9}$
5. 10   **6.** 30   **7.** $-\dfrac{21}{2}$ **8.** It has none.
9. $\dfrac{5}{7}$ **10.** It has none.   **11.** 3   **12.** $\dfrac{2}{3}$

■ Lesson 12.5

1. $x^4 - 12x^3 + 54x^2 - 108x + 81$
2. $t^3 + 12t^2 + 48t + 64$
3. $32x^5 + 80x^4 + 80x^3 + 40x^2 + 10x + 1$
4. $27 - 54x + 36x^2 - 8x^3$
5. $8a^3 - 12a^2b + 6ab^2 - b^3$
6. $27x^3 + 27x^2y + 9xy^2 + y^3$
7. $1 - 6x + 15x^2 - 20x^3 + 15x^4 - 6x^5 + x^6$
8. $x^4 - 12x^3y + 54x^2y^2 - 108xy^3 + 81y^4$
9. $x^7 + 14x^6 + 84x^5 + 280x^4 + 560x^3 + 672x^2 + 448x + 128$

■ Lesson 12.6

1. $1395.40   **2.** $349,100.78, $48,000
3. $848.87

■ Lesson 13.1

1. $\sin\theta = \frac{5}{13}$, $\tan\theta = \frac{5}{12}$, $\csc\theta = \frac{13}{5}$,
$\sec\theta = \frac{13}{12}$, $\cot\theta = \frac{12}{5}$

2. $\sin\theta = \frac{\sqrt{7}}{4}$, $\cos\theta = \frac{3}{4}$, $\tan\theta = \frac{\sqrt{7}}{3}$,
$\csc\theta \frac{4}{\sqrt{7}}$, $\cot\theta = \frac{3}{\sqrt{7}}$

3. $\sin\theta = \frac{3}{5}$, $\cos\theta = \frac{4}{5}$, $\csc\theta = \frac{5}{3}$,
$\sec\theta = \frac{5}{4}$, $\cot\theta = \frac{4}{3}$

4. $\cos\theta = \frac{\sqrt{3}}{2}$, $\tan\theta = \frac{\sqrt{3}}{3}$, $\csc\theta = 2$,
$\sec\theta = \frac{2\sqrt{3}}{3}$, $\cot\theta = \sqrt{3}$

5. $\sin\theta = \frac{2}{\sqrt{29}}$, $\cos\theta = \frac{5}{\sqrt{29}}$, $\tan\theta = \frac{2}{5}$,
$\csc\theta = \frac{\sqrt{29}}{2}$, $\sec\theta = \frac{\sqrt{29}}{5}$

6. $\sin\theta = \frac{1}{3}$, $\cos\theta = \frac{2\sqrt{2}}{3}$, $\tan\theta = \frac{\sqrt{2}}{4}$,
$\sec\theta = \frac{3\sqrt{2}}{4}$, $\cot\theta = 2\sqrt{2}$

7. $\sin\theta = \frac{\sqrt{11}}{6}$, $\cos\theta = \frac{5}{6}$, $\tan\theta = \frac{\sqrt{11}}{5}$,
$\csc\theta = \frac{6}{\sqrt{11}}$, $\cot\theta = \frac{5}{\sqrt{11}}$

8. $\sin\theta = \frac{\sqrt{15}}{4}$, $\tan\theta = \sqrt{15}$, $\csc\theta = \frac{4}{\sqrt{15}}$,
$\sec\theta = 4$, $\cot\theta = \frac{1}{\sqrt{15}}$

■ Lesson 13.2

1. $465°$, $-255°$ **2.** $285°$, $-435°$
3. $\frac{10\pi}{3}$, $-\frac{2\pi}{3}$ **4.** $\frac{7\pi}{4}$, $-\frac{9\pi}{4}$ **5.** $76°$, $166°$
6. $3°$, $93°$ **7.** $\frac{\pi}{6}$, $\frac{2\pi}{3}$ **8.** $\frac{\pi}{18}$, $\frac{5\pi}{9}$
9. $\frac{7\pi}{4}$ **10.** $-\frac{5\pi}{12}$ **11.** $\frac{\pi}{90}$
12. $\frac{37\pi}{180}$ **13.** $150°$ **14.** $-135°$
15. $40°$ **16.** $\approx 171.89°$

■ Lesson 13.3

1. -1 **2.** $-\frac{\sqrt{3}}{2}$ **3.** $-\frac{\sqrt{3}}{2}$ **4.** $\sqrt{2}$
5. $\sqrt{3}$ **6.** $\frac{2\sqrt{3}}{3}$ **7.** $\sqrt{3}$ **8.** $-\frac{\sqrt{2}}{2}$
9. $\frac{\sqrt{2}}{2}$ **10.** 2 **11.** $-\sqrt{3}$ **12.** -2

■ Lesson 13.4

1. $60° = \frac{\pi}{3}$ **2.** $45° = \frac{\pi}{4}$ **3.** $60° = \frac{\pi}{3}$
4. $-30° = -\frac{\pi}{6}$ **5.** $60° = \frac{\pi}{3}$ **6.** $45° = \frac{\pi}{4}$
7. $150° = \frac{5\pi}{6}$ **8.** $-30° = -\frac{\pi}{6}$
9. $135° = \frac{3\pi}{4}$ **10.** $30° = \frac{\pi}{6}$
11. $-45° = -\frac{\pi}{4}$ **12.** $120° = \frac{2\pi}{3}$

■ Lesson 13.5

1. $C = 100°$, $a \approx 4.8$, $b \approx 10.2$
2. $C = 100°$, $b \approx 25.8$, $c \approx 30.2$
3. $A = 38°$, $a \approx 22.0$, $c \approx 34.0$
4. No solution
5. $A \approx 40.9°$, $C \approx 84.1$, $c \approx 30.4$
6. $B \approx 71.8°$, $C \approx 78.2°$, $c \approx 39.2$;
$B \approx 108.2°$, $C \approx 41.8°$, $c \approx 26.7$

■ Lesson 13.6

1. $B \approx 74.5°$, $C \approx 43.5°$, $a \approx 51.3$
2. $A \approx 38.0°$, $C \approx 42°$, $b = 19.2$
3. $A \approx 38.2°$, $B \approx 99.8°$, $c = 23.8$
4. $A \approx 44.4°$, $B \approx 44.4°$, $C \approx 91.2°$
5. $A \approx 142.0°$, $B \approx 12.8°$, $C = 25.2°$
6. $A \approx 62.9°$, $B \approx 79.6°$, $C \approx 37.5°$

■ Lesson 14.1

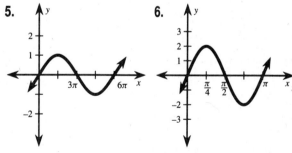

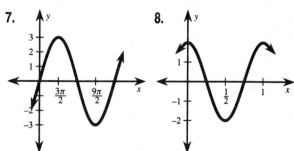

■ Lesson 14.2

1.
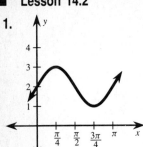

2.

3.
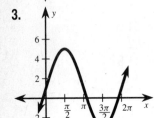

4.
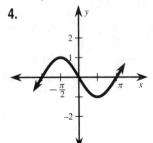

5.
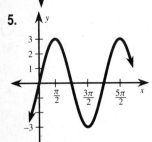

6.
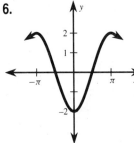

7.
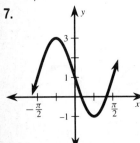

8.
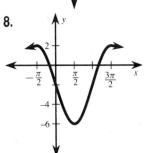

9.
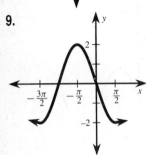

■ Lesson 14.3 (See page 113)

■ Lesson 14.4

1. $\frac{3\pi}{4}, \frac{7\pi}{4}$　　**2.** $0, \frac{\pi}{4}, \pi, \frac{7\pi}{4}$

3. $0, \frac{\pi}{6}, \pi, \frac{11\pi}{6}$　　**4.** $0, \frac{\pi}{2}, \pi, \frac{3\pi}{2}$

5. $\frac{\pi}{4}, \frac{3\pi}{4}, \frac{5\pi}{4}, \frac{7\pi}{4}$　　**6.** $\frac{\pi}{6}, \frac{5\pi}{6}, \frac{3\pi}{2}$

■ Lesson 14.5

1. $\frac{\sqrt{6}+\sqrt{2}}{4}$　**2.** $\frac{\sqrt{2}-\sqrt{6}}{4}$　**3.** $\frac{\sqrt{6}-\sqrt{2}}{4}$

4. $\sqrt{6}+\sqrt{2}$ or $\frac{4}{\sqrt{6}-\sqrt{2}}$　**5.** $\frac{\sqrt{6}+\sqrt{2}}{4}$

6. $\frac{\sqrt{3}-1}{\sqrt{3}+1}$　**7.** $\frac{\sqrt{6}-\sqrt{2}}{4}$　**8.** $\frac{-\sqrt{2}-\sqrt{6}}{4}$

■ Lesson 14.6

1. $\frac{\pi}{2}, \frac{4\pi}{3}, \frac{3\pi}{2}, \frac{5\pi}{3}$　　**2.** π

3. $\frac{\pi}{2}, \frac{7\pi}{6}, \frac{11\pi}{6}$　　**4.** $\frac{\pi}{6}, \frac{5\pi}{6}$

5. $\frac{3\sqrt{10}}{10}$ or $\frac{3}{\sqrt{10}}$　　**6.** $\sqrt{\frac{5}{6}}$ or $\frac{\sqrt{30}}{6}$

7. $-\sqrt{0.55}$　　**8.** $\frac{1}{3+\sqrt{8}}$ or $3-\sqrt{8}$

■ Lesson 15.1

1. $\frac{1}{5}$　**2.** 1　**3.** $\frac{1}{4}$　**4.** $\frac{501}{1847} \approx 0.271$

■ Lesson 15.2

1. 100,000　**2.** $9! = 362,880$

3. 6,400,000　**4.** $\frac{1}{7!} = \frac{1}{5040}$

■ Lesson 15.3

1. 56　**2.** $\frac{1}{15}$　**3.** 27,405　**4.** 300

■ Lesson 15.4

1. $\frac{9}{11}$　**2.** $\frac{1}{10}$　**3.** $\frac{5}{11}$　**4.** $\frac{4}{13}$　**5.** $\frac{12}{13}$

■ Lesson 15.5

1. $\frac{1}{216}$　　**2.** $\frac{1}{169}$

3. $\frac{1}{5^{20}} \approx \frac{1}{95,360,000,000,000}$　**4.** $\frac{4}{663}$

■ Lesson 15.6

1. \$3.15　**2.** $-\frac{1}{3} \approx -0.33$; no

3. $-\frac{4}{9} \approx -0.44$; no

Lesson 14.3

1. $\tan\left(\frac{\pi}{2} - x\right)\sin x = \cot x \sin x$
$$= \frac{\cos x}{\sin x} \cdot \sin x$$
$$= \cos x$$

2. $\frac{\tan^2 x}{\sec x} = \frac{\left(\frac{\sin x}{\cos x}\right)^2}{\frac{1}{\cos x}}$
$$= \frac{\sin^2 x}{\cos x}$$
$$= \frac{1 - \cos^2 x}{\cos x}$$
$$= \sec x - \cos x$$

3. $\frac{\sin x}{1 + \cos x} = \frac{\sin x(1 - \cos x)}{(1 + \cos x)(1 - \cos x)}$
$$= \frac{\sin x(1 - \cos x)}{1 - \cos^2 x}$$
$$= \frac{\sin x(1 - \cos x)}{\sin^2 x}$$
$$= \frac{1 - \cos x}{\sin x}$$

4. $\sec x \cot x = \frac{1}{\cos x} \cdot \frac{\cos x}{\sin x}$
$$= \frac{1}{\sin x}$$
$$= \csc x$$

5. $\cos^2 x(1 + \tan^2 x) = \cos^2 x \sec^2 x$
$$= \cos^2 x \cdot \frac{1}{\cos^2 x}$$
$$= 1$$

6. $\sin\left(\frac{\pi}{2} - x\right)\sec x = \cos x \cdot \frac{1}{\cos x}$
$$= 1$$

7. $\sec(-x)\cot(-x)\sin(-x) = \sec x(-\cot x)(-\sin x)$
$$= \frac{1}{\cos x} \cdot \frac{\cos x}{\sin x} \cdot \sin x$$
$$= 1$$

8. $\cos x(\csc x + \tan x) = \cos x\left(\frac{1}{\sin x} + \frac{\sin x}{\cos x}\right)$
$$= \frac{\cos x(\cos x + \sin^2 x)}{\sin x \cdot \cos x}$$
$$= \frac{\cos x}{\sin x} + \sin x$$
$$= \cot x + \sin x$$